C# Fundamentals – Getting Started with C# 11 and .NET 7

Adam Seebeck

United States

C# Fundamentals – Getting Started with C# 11 and .NET 7

Author: Adam Seebeck

Copy Editor: Samantha Seebeck

For permission to use material from this text or product, please contact www.unQbd.com/Support or email PermissionRequest@unQbd.com

Paperback ISBN: 978-1-954086-15-9
Hardcover ISBN: 978-1-954086-14-2
Kindle ISBN: 978-1-954086-12-8

Edition: 5.0

unQbd
100 S. Belcher #5483
Clearwater, FL 33765
USA

Print year: 2022

Unless otherwise noted all items © unQbd.

unQbd (pronounced "un-cubed") is a streaming service that offers a variety of books that can be read on your choice of internet-connected device. You can read as much as you want for one low yearly price. unQbd books are also available for purchase in print form.

Unlimited books, audio books, and more.

Membership benefits

Your free, 7-day trial comes with:

Unlimited access

 Full access to all books.

 Full access to all audio books.

Community access

 Socialize with other members and *Authors*.

 Ask a question directly within an interactive book.

 Share your thoughts on books.

Trial Reminder

 A friendly email reminder before your trial ends.

Start your membership today at unQbd.com

C# Fundamentals: Getting Started with C# 11 and .NET 7

ABOUT THIS BOOK

C# Fundamentals explains essential software development concepts and provides easy to follow examples. This book was created by a higher education instructor and is structured as a course that includes activities and quizzes to help practice what you learn throughout the book. This book utilizes the following five teaching techniques:

Quick Reference Essentials give a brief explanation and include examples of code. Every section in the book begins this way to quickly convey the overall topic that is being covered.

Full Examples often follow Quick Reference Essentials and expand on the concept by displaying complete real-world examples of usage.

Walkthroughs provide step-by-step instructions on how to complete a specific task. The Walkthroughs contain all information n eeded to complete a fully functional application.

Activities are assignments that can be completed to practice a concept. Activities range in complexity but are based on the information learned in the section.

Mini Quizzes are a quick way to gauge comprehension of a given topic.

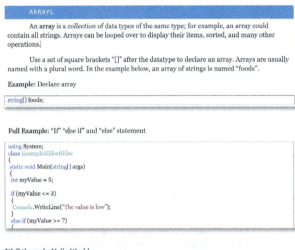

This book focuses on software development using the C# language and computer software called **Visual Studio** (free to download). Both the language and software are covered within this book ; no prior software development knowledge is required. Flowcharts, screenshots, and additional visual aids are used throughout the book to help explain concepts.

INTRODUCTION TO C#

C# (pronounced "C Sharp") is a modern, easy to use, **object-oriented** programming language developed by the Microsoft Corporation. The first version of C# was released in 2002 and the latest version (11.0) was released in November 2022. C# is the most popular language used within Microsoft's Visual Studio platform. Visual Studio is the software used for developing applications; this type of software is referred to as an **Integrated Development Environment (IDE)**.

C# uses the **.NET Platform**, which contains all the components needed to run an application. One component is the **Common Language Runtime** (CLR). The CLR handles various tasks, such as compiling code to 1s and 0s (binary) for the computer to understand. The new **.NET 7** framework is also within the .NET Platform and contains shared libraries of code that developers use when developing applications. This helps developers by not have to write code from scratch. For example, if an application needs to read information from a text file, there is already a built-in class for this and there is no need to recreate it

An additional benefit to using C# is its versatility. C# is a general purpose language, meaning it can be used to create many kinds of applications.

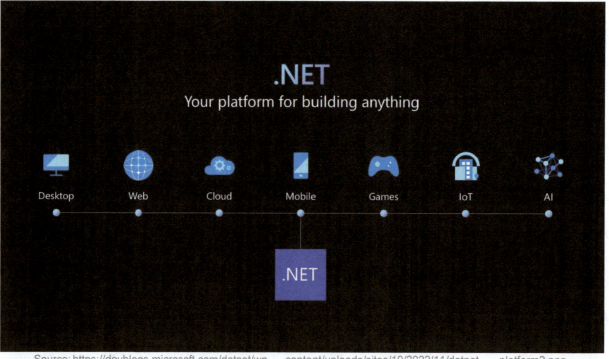

Source: https://devblogs.microsoft.com/dotnet/wp   -content/uploads/sites/10/2022/11/dotnet   -platform2.png

Listed below are a few examples of apps that can be created using the C# language.

Console Applications: A console application is a text-based application that contains no graphical components, such as buttons or images. This book focuses on text-based applications because they easily demonstrate C# examples.

Websites: Dynamic websites can be built using C# and ASP.NET in Visual Studio.

Mobile Applications : Quickly create iOS and Android applications using C# and .NET MAUI in Visual Studio.

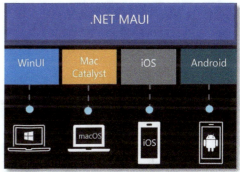

Games Unity is a separate (free) software program used to design 2D, 3D, virtual, and mixed reality games. While the Unity Game Engine is used to design the game, the coding is still developed in Visual Studio with C#.

Computer Software: Software for Windows can be created using WPF, Windows Forms, or .NET MAUI.

VISUAL STUDIO: INSTALLING

Visual Studio can be downloaded from "www.VisualStudio.com/downloads". Download the Community edition; it is free.

The screen below will be displayed after downloading and beginning the installation process. This screen contains the components that can be installed. Components can always be added later by going to the top menu bar and selecting "Tools -> Get Tools and Features".

For this book, make sure to have at least ".NET desktop development" checked. Review all the additional components available to install to become familiar with the options. Click the install or modify button to continue the installation process.

VISUAL STUDIO: CONSOLE APPLICATION SETUP

The following instructions explain how to create a **Console Application**. Console Applications are great for demonstrating concepts; they are simple text-based applications.

Walkthrough : Microsoft Windows setup for a Console Application ***may vary for Mac computers*

1. Open *Visual Studio* from your Desktop or Start Menu.

2. Click "Create a new project".

3. In the Project types dropdown select "Console" and select "Console App".

4. Click "Next" in the bottom right corner .

5. Give the project a name, select a location to save the project, and then click "Next" in the bottom right corner.

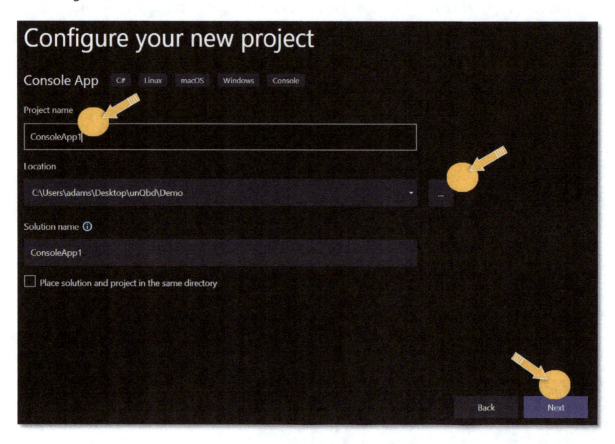

6. Click the "Create" button in the bottom right corner.

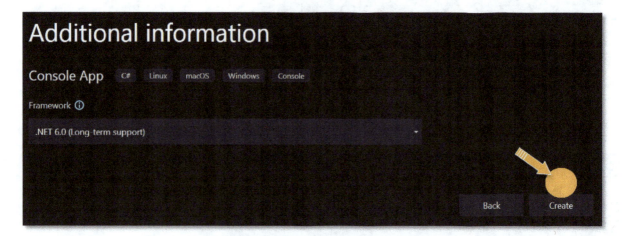

CONSOLE APPLICATION: "HELLO WORLD!"

When learning a new programming language, it is a tradition to display the words "Hello, World!" on the screen. Please note, this is the only example in the book where an explanation is not provided about the coding and simply the instructions are given to produce the results.

Walkthrough : Hello World

1. Crete a Console application named *HelloWorld* . The steps to create a Console application can be found in the previous section "*Visual Studio: Console Application Setup*".

2. By default, Console applications display "Hello, World!" when created. Please note, the file "Program.cs" should automatically load, if it does not double click it from the Solution Explorer menu.

3. Press the play button or the F5 key to run the Console application .

4. The "Hello, World" Console application will load.

Congratulations on completing a C# program!

COMMENTS

Program **comments** are notes left on programs to explain or highlight the functionality of code. Comments are **non-executing code** and will not affect the usage of a program.

Comment Types:

- **Line comments** are for a single line of code and begin with two forward slashes "//". Anything following the slashes on that line of code will be considered a comment and not execute.
- **Block comments** are for multiple lines of code and begin with one forward slash and an asterisk "/*" and to complete the comment end with an asterisk and a forward slash "*/".

Example: Line comment

```
// Anything after two forward slashes on this line will not execute.
Console.Write("Hello World!"); // A comment can also be placed after code.
```

Example: Block comment

```
/* Comment out multiple lines of code.
  Anything in this area is also a comment.
*/
```

Visual Studio has a convenient way to quickly comment or uncomment selected lines of code. Click on the Comment or Uncomment buttons.

SNIPPET & SHORTCUT KEYS

Snippet and **shortcut keys** can help save time while coding in Visual Studio. With snippet keys, enter only a few letters or a keyword and then press tab twice to insert a code fragment.

Example: cw (tab) (tab)

```
// cw (tab)(tab) : Type cw and then press the tab key twice, the line below will display
Console.WriteLine();
```

Snippet & Shortcut Keys	Result
cw (tab) (tab)	Console.WriteLine();

while (tab) (tab)	while (true)
	{
	}

for (tab) (tab)	for (int i = 0; i < length; i++)
	{
	}

foreach (tab) (tab)	foreach (var item in collection)
	{
	}

do (tab) (tab)	do { } while (true);
prop (tab) (tab)	public int MyProperty { get; set; }
(F5)	Will start debugging
(ctrl) + (F5)	Will start without debugging

VARIABLES AND DATA TYPES

A **variable** is a named location in computer memory that holds information for later use. The information within the variable can change or can be used when needed. Think of a variable as a box that has been named; the contents within the box can change, but the box name stays the same.

The contents within a variable can have different **data types**. The data type declares what type of information is going to be contained within the variable. There are many kinds of data types, such as a *string* for text ("Hello World") or an *integer* for numbers (54) with which mathematical calculations can be performed.

In the example below, a variable of data type string is declared with a variable name of "myStringName". Declaring a variable can be thought of as creating an *empty* box that is named.

Example: Declare a variable

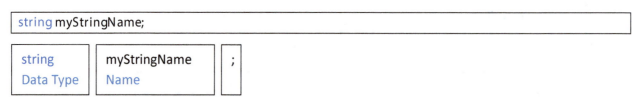

Example: Declare a variable *and* assign a value

The variable "myStringName" now contains the value "Hello" within it. In the example below, the value is changed to "Hello again" and the data type is not declared again when changing the value. Once the data type is declared, it is not declared again.

Example: Assign a new value

```
myStringName = "Hello again";
```

Example: Declare a variable and assign an integer value

```
int myIntName = 2 + 3; // myIntName is now 5
```

A string puts the two values together; an integer *adds* them together.

When declaring certain numeric data types, such as a *float* or a *decimal*, it is required to put a character behind the value. The example below demonstrates this.

Example: Numeric data types

```
int numOne = 5;
float numTwo = 2.55f; //  When declaring a float an "f" needs to be after the value
double numThree = 3.33;
decimal numFour = 4.66m; //  When declaring a decimal an "m" needs to be after the value
```

It is recommended to be specific when declaring data types to avoid confusion. However, it is possible to use the data type **var** and the compiler will attempt to figure out which data type to use.

Example: var

```
var theWord = "hello"; // Declares a string and assigns value "hello"
var theNumber = 5;     // Declares an int and assigns value 5
```

Every data type uses a specific amount of memory when declared. The smallest measurement is a **bit**, and eight bits make up a **byte**. The chart below lists a few of the most common data types along with how many bytes they use, their range, and an example of their data.

Data Types	Bytes	Range	Example
byte	1	0 to 255	7
short	2	-32,768 to 32,767	-10
int	4	-2,147,483,649 to 2,147,483,647	12
long	8	-9,233,372,036,854,775,808 to 9,233,373,036,854,775,807	-54
sbyte	1	-128 to 127	5
ushort	2	0 to 65,535	7
uint	4	0 to 4,294,967,295	9
ulong	8	0 to 18,446,744,073,709,551,615	11
float	4	Represents a floating point value up to 7 digits	12.3
double	8	Represents a floating point value up to 16 digits	12.35
decimal	16	Represents a floating point value up to 29 significant digits	12.356
bool	1	Logical Boolean type (can only be True or False)	True
char	2	A single Unicode character	H
string	varies	A sequence of characters	Hello

Integral types can be **signed types** or **unsigned types**. A signed type value can be positive or negative and unsigned types can only be positive values. For example, a short, which is a signed type, can have a value from -32,768 to 32,767. However, if negative numbers are not included, then the positive numbers will double when making it unsigned. Using the unsigned data type for short (which is ushort), the range will become 0 to 65,535.

MINI QUIZ: VARIABLES AND DATA TYPES

Mini Quiz

1. Which of the following is a data type?

    A. 5
    B. Int
    C. "Hello"
    D. Console.ReadLine()

2. What data type is the best to use if a number from 1 to 10 need to be stored?

    A. Int
    B. String
    C. Byte
    D. Char

3. Will the following code give an error message?

```
int myNumber = "5";
```

    A. Yes
    B. No

4. What character needs to be placed after a decimal value when being declared?

    A. M
    B. F
    C. D
    D. A character is not needed

5. Can an unsigned type have the value 21?

    A. Yes
    B. No

Mini Quiz Answers

1. B
2. C – Bytes have a range from 0 to 255 and will use the smallest amount of memory
3. A – Needs to be 5 not "5"
4. A
5. B – Unsigned types can not be negative

NAMING CONVENTIONS

Using proper naming and capitalization techniques can greatly increase the readability of code. In C# the two most popular capitalization techniques for naming variables, classes, methods, etc. are **Camel Casing** and **Pascal Casing**

- **Camel Casing** The first letter of the first word is lowercase, and all subsequent words start with a capital letter.
- **Pascal Casing** The first letter of every word is capitalized.

Example: Camel Case

```
int myFavoriteNumber = 7;
```

Use Camel Casing for local variables and arguments. Pascal Casing is used in most other situations. While moving ahead in this book take notice of which casing technique is used.

Example: Pascal Case

```
public void MyMethodName() { }
```

Names should not include hyphens, spaces, non-alphanumeric characters, and they do not start with a numeric number. The name should also be **self-documenting**. For example, instead of calling a variable "variable1", which has no meaning, it should be "scoreTotal", which describes what the variable is.

MINI QUIZ: NAMING CONVENTIONS

Mini Quiz

1. Select all the valid Camel Casing names.

    A. response
    B. my Response
    C. 2ndResponse
    D. userResponse
    E. my-Response

2. Select all the valid Pascal Casing names.

    A. response
    B. My Response
    C. 2NdResponse
    D. UserResponse
    E. My-Response

3. Which of the following would NOT be considered a self-documenting name?

    A. Method1
    B. totalScore
    C. HealthLevel

Mini Quiz Answers

1. A, D – Incorrect: B has a space, C starts with a number, and E has a dash
2. D – Incorrect: A starts with a lower-case letter, B has a space, C starts with a number, and E has a dash
3. A – Does not describe what the method does.

OPERATORS

The following is a sample of common **operators** used in C#.

Arithmetic	+ - * / %	Arithmetic operators perform a *mathematical* function. + Add - Subtract * Multiply / Divide % Remainder **Example:** 5 + 1 // Output: 6
Assignment	= += -= *= /= %=	Assignment operators *assign* a value **Example:** X = 5 // X is assigned the value 5
Relational	== != < > <= >=	Relational operators are used to *compare* values == Equality (equal to) != Inequality (not equal to) < Less than > Greater than <= Less than or equal >= Greater than or equal **Example:** X == 5 // Compares to see if X is equal to 5
Increment /Decrement	++ --	Increases or decreases a value **Example:** X++ // Increases the value by 1
String Concatenation	+	The plus sign will combine the items (not mathematically) **Example:** 5 + 4 // Output: 54 **Example:** "Hi" + " John" // Output: Hi John
Conditional	&& ||	&& checks if both conditions are true **Example:** (5 == X) && (6 == Y) || checks if either condition is true **Example:** (5 == X) || (6 == Y)

INPUT & OUTPUT

Console applications are great for demonstrating computer concepts, which is why this book focuses on them. They can easily **output** text and accept user **input** .

"Write" and "Write Line" can both be used to output text in a Console application. The example below demonstrates how to enter text between the quotes for it to display on the console.

Example: Output text

```
Console.Write ("Hello ");       // Write does not append a new line when finished
Console.Write(" World! ");      // Output: Hello  World!

Console.WriteLine("Hello ");    // WriteLine appends a new line when finished
Console.WriteLine(" World! ");  // Output: Hello
                                //          World!
```

Below is a brief description of what each part of the above code represents.

```
Console.WriteLine("Hello World!");
```

Console. Selects the Console Class (Classes covered in a later section)	WriteLine Tells the console to output text	("	Hello World! The text to output on the console	");

"ReadLine" is used to input text in a Console application. "ReadLine" waits until text is entered in the Console application and for the "Enter" key to be pressed.

Example: Input text

```
Console.ReadLine();
```

Example: Assign user input to a variable

```
string saveToVariable = Console.ReadLine();
```

User input from "Console.ReadLine()" is *always* a string. If the variable needs to be an integer, for example, you would need to convert it from string to int (covered later in the "Parse and TryParse" section).

There is a way to shorten the code from having to type "Console.WriteLine" to just "WriteLine". Add "using static System.Console;" to the very top of Program.cs. A more detailed explanation on this topic is covered later in "Namespaces & Using Directives".

Example: WriteLine and ReadLine without using " Console."

```
using static System.Console;"

WriteLine(" Console. is no longer needed!");
ReadLine();
```

There are several different techniques to output data. In the examples below, the same information is displayed in multiple ways. The last example uses the dollar sign ($) before the quotes. This denotes the usage of **string interpolation** . String interpolation helps provide a more readable and convenient syntax and is the format this book will use going forward.

Full Example: Variables displayed in multiple ways

```
string todaysDay = "Tuesday";

Console.WriteLine(todaysDay);                    // Output: Tuesday
Console.WriteLine( "Today is " + todaysDay);   // Output:  Today is Tuesday
Console.WriteLine( "Today is {0}", todaysDay); // Output: Today is Tuesday
Console.WriteLine( $"Today is {todaysDay}");   // Output: Today is Tuesday
```

Escape characters are a special sequence of characters that are used to display something other than the exact interpretation. For example, entering in quotes within "Console.Writeline (" "Hello" ")" quotes will cause an issue To get around this issue, use a backslash.

Example: Escape character for quote

```
Console.WriteLine( "Add a backslash before a quote \"Hello\"");
// Output: Add a backslash before a quote "Hello"
```

Escape characters can also be useful for starting a new line ("\n"), tab ("\t"), and many other options.

Example: Tab and Newline

```
Console.WriteLine( "\t Hello \n World" );
// Output:     Hello
//             World
```

The "@" symbol can be used before text to make the entire string ignore escape characters; the technical name for this is **verbatim string literal** . This is often helpful with file names that would require many double backslashes ("\\").

Example: File name with backslashes compared to file name with "@" symbol

```
Console.WriteLine( "C:\\Users\\ComputerUser\\Desktop\\AFile.doc"); // with backslashes
Console.WriteLine( @"C:\Users\ComputerUser\Desktop\AFile.doc");    // with @ symbol
```

ACTIVITY: DISPLAY USER INPUT

Activity: This activity focuses on asking a user for information and then repeating it back. Create a Console Application that meets the three requirements listed below.

1. Display the question "Hello, what is your name?".

2. Save the users response to a variable.

3. Output the response "Thank you, {name that was saved}".

The finished output of the program should look like the example below:

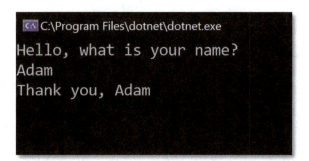

Need help with the solution? https://www.unQbd.com/Solutions/CSharp5th/DisplayUserInput

CASTING

The conversion from one type to another is called **typecasting**, usually referred to as **casting**. The two types of casting are **implicit casting** and **explicit casting**.

- **Implicit casting** occurs without specific instructions saying to cast. In most cases this will occur automatically when going from a *narrow* type to a *wider* type with no loss of data.
- **Explicit casting** occurs because the instructions stated for the casting to happen. Going from a *wider* type to a *narrow* type generally requires explicit casting to acknowledge there may be loss of data.

When adding numbers together they both must be the same type. In the example below a short and an int are added together using *implicit* casting. There is no chance of data loss because a short is going from a narrow type to a wider type. A short is numbers -32,768 to 32,767, and an int is numbers -2,147,483,649 to 2,147,483,647.

Example: Implicit casting

```
short num1 = 50;
int num2 = 600;
int sum = num1 + num2; // num1 is implicitly cast to int
```

Example: Explicit casting

```
decimal firstNumber = 50.1234m;
int secondNumber= 2;

// Explicitly cast firstNumber to int. Data was lost (.1234)
int sum = (int)firstNumber + secondNumber; // Output: 52
```

Casting is ideal for going from similar data types, such as short to int. However, if you need to convert from string to an int, that is covered in the upcoming "Parse and TryParse" section in this book.

MINI QUIZ: CASTING

Mini Quiz

1. What type of casting is being used in the following example?

```
long firstNumber = 603453;
int secondNumber = 54;
int total = (int)firstNumber + secondNumber;
```

    A. Implicit
    B. Explicit
    C. Will throw an error

2. What type of casting is being used in the following example?

```
byte firstNumber = 60;
int secondNumber = 700;
int total = firstNumber + secondNumber;
```

    A. Implicit
    B. Explicit
    C. Will throw an error

3. What type of casting is being used in the following example?

```
long firstNumber = 12;
int secondNumber = 343;
int sum = firstNumber + secondNumber;
```

    A. Implicit
    B. Explicit
    C. Will throw an error

Mini Quiz Answers

1. B
2. A
3. C

IF, ELSE IF, AND ELSE STATEMENTS

Decision-making is an important part of computer programs. The "**if**", "**else if**", and "**else**" statements identify which statement to run based on the value of a **Boolean** (true or false) expression. The visual example below asks the question "Is today Monday?"; it must have a true or false answer.

The code within an "if" statement will only execute if the condition is *true*. In the example below there is an "if" statement with an **equality operator** which is "==". In a previous section called "Operators", there is a list of relational operators that can be used with in "if" statements.

Example: "If" statement

```
if ("Monday" == "Monday")
{
 // Code within the curly brackets will execute because Monday equals Monday is TRUE
}
```

"If" statements should contain at least one variable. In the example above, Monday will always equal Monday, so there is no reason to use an "If" statement.

Example: "If" statement with a variable

```
string todaysDay ="Monday";

if (todaysDay == "Monday")
{
 // Code within the curly brackets will execute because Monday equals Monday is TRUE
}
```

Using an "else" statement *after* an "if" statement is a catch all. Meaning that if the "if" statement is not true, execute the code within the "else" statement.

Example: "If" and "else" statement

```
string todaysDay ="Tuesday";

if (todaysDay == "Monday") // This will  NOT execute, todaysDay does not equal Monday
{
 Console.WriteLine( "Today is Monday");
}
else // This  WILL  execute, the previous if statement was not true
{
 Console.WriteLine( "Today is NOT Monday"); //  Output: "Today is NOT Monday"
}
```

Use an "else if" statement *after* an "if" statement to test additional conditions to be true. Only one "if" statement and one "else" statement can be used However, multiple "else if" statements are allowed. As soon as an "if" or "else if" statement is true, only the code within that code block is executed. Any additional statements following that one will not execute even if they are true. The order in which the "else if" statements is placed is important.

Example: "If" "else if" and "else" statement

```
string todaysDay ="Tuesday";

if (todaysDay == "Monday")               // This will not execute, todaysDay does not equal Monday
{
 Console.WriteLine( "Today is Monday");
}
else if (todaysDay == "Tuesday")         // This  WILL  execute, todaysDay equals Tuesday
{
 Console.WriteLine( "Today is Tuesday"); // Output: "Today is Tuesday"
}
else if (todaysDay == "Wednesday")       // This will not execute, a previous If  statement was true
{
 Console.WriteLine( "Today is Wednesday");
}
else                                     //  This will not execute, a previous If statement was true
{
 Console.WriteLine( "Today is NOT Monday, Tuesday, or Wednesday");
}
```

The curly brackets following a condition are not always needed. If the statement is exactly one line of code they are not needed. They are only required for multiple lines of code. This book always uses curly brackets (excluding example below), although they are not required.

Example: Curly brackets

```
string todaysDay = "Tuesday";

if (todaysDay == "Monday")
 Console.WriteLine( "Today is Monday");
else if (todaysDay == "Tuesday")
 Console.WriteLine( "Today is Tuesday"); // Output: "Today is Tuesday"
else if (todaysDay == "Wednesday")
{
 // Curly brackets are needed if there are more than 1 statement following a condition
 Console.WriteLine( "Today is Wednesday");
 Console.WriteLine( "It is the middle of the work week!" );
}
else
 Console.WriteLine( "Today is NOT Monday, Tuesday, or Wednesday");
```

Full Example: "If" "else if" and "else" statement

```
int myValue = 5;

if (myValue <= 3)
  {
  Console.WriteLine( "The value is low");
  }
else if (myValue >= 7)
  {
  Console.WriteLine( "The value is high");
  }
else // The else value is true because the first 2 conditions were not met
  {
  Console.WriteLine( "The value is in the middle"); // Output: The value is in the middle
  }
```

"If" statements can be **nested**, meaning they can be put inside other "if" statements.

Example: Nested statement

```csharp
bool gameActive = true;
int health = 40;

if (gameActive == true)
{
  Console.WriteLine("Game is Active!");

 if (health == 100) // Nested "if" statement
 {
  Console.WriteLine("You are in perfect health!");
 }
 else
 {
  Console.WriteLine("You have been hurt");
 }
}
```

Conditional operators (covered in the "Operators" section) are commonly used in "if" and "else if" statements. In the example below the "&&" operator is used, meaning that both conditions must be true.

Example: Conditional operator "AND" && (both conditions must be true)

```csharp
bool gameActive = true;
string player = "Adam";

if (gameActive == true && player == "Adam")
{
 Console.WriteLine("The game is active AND Adam is alive!" );
}
```

Example: Conditional operator "OR" || (either condition can be true)

```csharp
bool gameActive = true;
string player = "Adam";

if (gameActive == true || player == "Adam")
{
 Console.WriteLine("The game is active OR Adam is Alive!");
}
```

ACTIVITY: MAKING DECISIONS

Activity: This activity focuses on making decisions based on your mood. Create a Console Application that meets the three requirements listed below.

1. Display the question "What kind of mood are you in? Good, Ok, or Bad?".

2. Save the users response to a variable.

3. Have different responses based on what they typed in. Make sure to have a response if a user does not type in one of those three options.

The finished output of the program should look like the examples below:

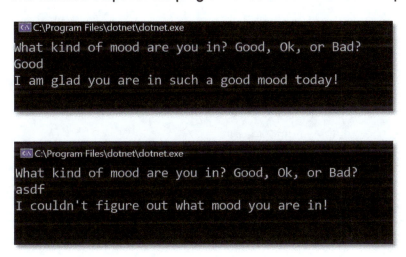

Need help with the solution? https://www.unQbd.com/Solutions/CSharp5th/MakingDecisions

SWITCH

The **switch** statement is like an "if" statement; it is useful if you have several options from which to choose. Anything that can be done with a switch statement can also be done with an "if" statement. Select whichever makes the code easiest to understand.

With switch statements each option is a **case** and there is also a **default** option if none of the cases match (like an "else" statement). Within the case, curly brackets "{}" are not needed. A case is ended with "break;".

Example: Switch

```
int optionPicked = 3;

switch (optionPicked)
{
case 1:
 Console.WriteLine("My number is 1");
 break;
case 2:
 Console.WriteLine("My number is 2");
 break;
case 3: // case 3 is true
 Console.WriteLine("My number is 3"); // Output: My number is 3
 break;
default:
 Console.WriteLine("Other option");
 break;
}
```

If there is code within a case, "break" must be used. If the code within a case is empty, the case will "fall through" until a "break" is reached. In the example below several cases are combined into one result.

Example: Switch Case "fall through"

```csharp
int optionPicked = 2;

switch (optionPicked)
{
 case 1:
 case 2:
 case 3:
 // The text below will display
 Console.WriteLine( "Low number selected"); // Output: "Low number selected"
 break;
 case 4:
 case 5:
 Console.WriteLine( "Medium number selected");
 break;
 default:
 Console.WriteLine( "Other number selected");
 break;
}
```

Full Example: Switch

```csharp
Console.WriteLine( "Type the name of a number: One, Two, Three, or Four");

string num1 = Console.ReadLine();

switch (num1)
{
 case "One":
 case "Two":
 Console.WriteLine( "Picked number 1 or 2");
 break;
 case "Three":
 Console.WriteLine( "Picked number 3");
 break;
 case "Four":
 Console.WriteLine( "Picked number 4");
 break;
 default:
 Console.WriteLine( "Entered incorrect input" );
 break;
}
```

ACTIVITY: SWITCH

Activity: This activity focuses on using a switch statement to provide feedback based on what letter is entered. Create a Console Application that meets the two requirements listed below.

1. Display the following "Enter a letter and find out if it is a vowel".

2. Use a switch statement to provide the following 3 responses for the letter entered.

    A. The lowercase {letter} is a vowel
    B. The uppercase {letter} is a vowel
    C. The {letter} is NOT a vowel

The finished output of the program should look like the examples below:

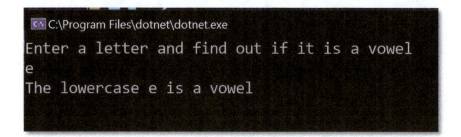

Need help with the solution? https://www.unQbd.com/Solutions/CSharp5th/Switch

PARSE & TRYPARSE

Parse and **TryParse** both attempt to convert one data type to another. The example below attempts to convert a *string* to an *int*. When using Parse, be 100% certain that it will convert successfully, otherwise the program will crash. For example, if a user is asked to input a numeric number and they type the word "Hello", the program will crash.

TryParse is a safer approach because it checks if the value can convert to a numeric value. If TryParse cannot convert the value, it sets the value to 0.

Example: Parse from a string to a numerical value

```
int exampleOneInt = int.Parse('6");              // Parse a string to an int

string exampleTwo = "7";
int exampleTwoInt = int.Parse(exampleTwo);       // Parse a string to an int

string exampleThree = "6.5";
double exampleThreeInt = double.Parse(exampleThree);// Parse a string to a double
```

The above examples work, however if any value is entered incorrectly the program would crash. For example, if exampleTwo = "Hello" or "Seven", the program would crash.

Example: TryParse from a string to an int

```
string textExample = "7";
int textExampleInt;
int.TryParse(textExample, out textExampleInt);
```

In the example above, if textExample = "Hello" or Seven", the program would display 0 by default and NOT crash.

Example: TryParse using an **Inline** Out variable

```
string textExample = "10";
int.TryParse(textExample, out int textExampleInt);
```

Using the Inline Out variable in the approach above, it is not needed to declare the textExampleInt int beforehand.

Example: TryParse using an "If" statement

```
string textExample = "8";
if (int.TryParse(textExample, out int textExampleInt))
{
 Console.WriteLine($"TryParse was successful. Converted a String to an Int: {textExampleInt}");
}
else
{
 Console.WriteLine("TryParse failed. textExampleInt set to the default which is 0.");
}
```

Full Example: Parse

```
Console.WriteLine("Enter a number");
string text1 = Console.ReadLine();
// This is unsafe, because if the user enters incorrect information the
// program will crash. Use TryParse instead.
int num1 = int.Parse(text1);

Console.WriteLine("Enter a second number");
// ReadLine can be used directly in Parse to accept user input
// This is unsafe because if the user enters incorrect information the
// program will crash. Use TryParse instead.
int num2 = int.Parse(Console.ReadLine());

Console.WriteLine($"The sum of num1: {num1} and num2: {num2} is {num1 + num2}");

// Parse below is safe because there is a 100% chance 6 will convert into an int
string text3 = "6";
int num3 = int.Parse(text3);
```

Full Example: TryParse

```
Console.WriteLine("Enter a number");
string text1 = Console.ReadLine();

int.TryParse(text1, out int num1);

Console.WriteLine("Enter a second number");
// ReadLine can be used directly in the TryParse to accept user input
int.TryParse(Console.ReadLine(), out int num2);

Console.WriteLine($"The sum of num1: {num1} and num2 {num2} is {num1 + num2}");
```

ACTIVITY: TRYPARSE

Activity: This activity focuses on converting from one data type to another. Create a Console Application for a bill calculator that meets the three requirements listed below.

1. Display the following: "Bill Calculator".

2. Ask the three questions below and accept user input for each question.

    A. How much was the entrée?
    B. How much was the drink?
    C. How much was the dessert?

3. Add all three of the values and provide the total.

The finished output of the program should look like the example below:

```
C:\Program Files\dotnet\dotnet.exe
Bill Calculator
How much was the entree? 17.99
How much was the drink? 1.45
How much was the dessert? 4.99
The total for all 3 items is: $24.43
```

Need help with the solution? https://www.unQbd.com/Solutions/CSharp5th/TryParse

LOOPS

A **Loop** can repeat a block of code multiple times. The four main types of loops are **While**, **Do-While**, **For**, and **ForEach**.

A **While Loop** executes a block of code while the condition is *true*.

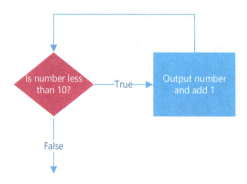

Example: While Loop

```
int number = 0;

while (number < 10)
{
 Console.Write(number);
 number++; // number++ is equivalent to using "number = number + 1"
}
 // Output: 0123456789
```

The above code will loop and output numbers from 0 to 9. Every time the loop executes it increases the value by 1 until the number is equal to 10.

A **Do-While Loop** evaluates the condition *after* the loop has executed, meaning the code block will always execute at least once.

Example: Do-While Loop

```
int number = 0;

do
{
 Console.Write(number);
 number++;
}
while (number < 10);

// Output: 0123456789
```

A **For Loop** is generally used when the number of iterations needed is known. The starting value, test conditions, and the increment/decrement are all set in one compact line of code.

Example: For Loop

```
int number = 10;

for (int i = 0; i < number; i++)
{
 Console.Write(i); // Output: 0123456789
}
```

Curly brackets {} are not needed if a loop is only 1 line of code. However, this book always uses curly brackets.

A **ForEach Loop** is used for collections of items, such as arrays (arrays covered in next section).

Example: ForEach Loop

```
int[] numbers = new int[10] { 0, 1, 2, 3, 4, 5, 6, 7, 8, 9 };

foreach (int number in numbers)
{
 Console.Write(number); // Output: 0123456789
}
```

Full Example: While and For loops

```
Console.WriteLine("Enter a number to count up to it using a While loop");

int.TryParse(Console.ReadLine(), out int num1);

int counter = 0;

while (counter <= num1)
{
  Console.WriteLine(counter);
  counter++; // number++ is equivalent to using "number = number + 1"
}

  Console.WriteLine("Enter a number to count up to it using a For loop");
  int.TryParse(Console.ReadLine(), out int num2);

for (int i = 0; i <= num2; i++)
{
  Console.WriteLine(i);
}
```

Putting a loop within another loop is called a **nested loop**. The outer loop is called the **parent loop**, and the loop within is the **child loop**. In the example below, the outer loop controls when to exit the game.

Example: Nested loop

```csharp
string playGame = "";
int counter = 0;

while (playGame != "Quit") // "!=" means not equal to
{
 while (counter < 5)
 {
   counter++;
   Console.WriteLine($"This number is from the nested loop {counter}");
 }

 counter = 0;
 Console.WriteLine("Press Enter to display again or type Quit to exit.");
 playGame = Console.ReadLine();
}
```

Exit a loop early by using the **break** keyword. In the example below, the program is ended when the user guesses the correct number.

Example: Break

```csharp
string computerNumber = "5";

while(true) // Generally do not use a forever loop; this was used as an example for break
 {
   Console.WriteLine("Guess what number I am thinking");

   string guessedNumber = Console.ReadLine();

   if (guessedNumber == computerNumber)
  {
     break;
  }
 }
```

There is another keyword like break called **continue**. Instead of exiting the loop, continue goes directly to the start of the loop to check the condition again. In the example below, continue is used to skip the number 7.

Example: Continue

```csharp
for (int i = 0; i < 10; i++)
{
 if (i == 7) // Skip number 7
 {
 continue;
 }
 Console.WriteLine($"Number: {i}");
}
```

ACTIVITY: LOOPS

Activity: In this activity a number is entered and then will count up or down to that number. Create a Console Application for a Counter App that meets the three requirements listed below.

1. Display the following: "Counter App".

2. Display the two prompts below and accept user input for each.

    A. Enter in a number:
    B. Do you want to count Up or Down from that number? (Options: Up or Down):

3. Display the output of either counting up or down to that number.

The finished output of the program should look like the examples below:

```
C:\Program Files\dotnet\dotnet.exe
Counter App
Enter in a number: 7
Do you want to count Up or Down from that number? (Options: Up or Down): Up
Counting Up: 0
Counting Up: 1
Counting Up: 2
Counting Up: 3
Counting Up: 4
Counting Up: 5
Counting Up: 6
Counting Up: 7
```

```
C:\Program Files\dotnet\dotnet.exe
Counter App
Enter in a number: 8
Do you want to count Up or Down from that number? (Options: Up or Down): Down
Counting Down: 8
Counting Down: 7
Counting Down: 6
Counting Down: 5
Counting Down: 4
Counting Down: 3
Counting Down: 2
Counting Down: 1
Counting Down: 0
```

Need help with the solution? https://www.unQbd.com/Solutions/CSharp5th/Loops

ARRAYS

An **array** is a *collection* of data types of the same type; for example, an array could contain all strings. Arrays can be looped over to display their items, sorted, and many other operations.

Use a set of square brackets "[]" after the datatype to declare an array. Arrays are usually named with a plural word. In the example below, an array of strings is named "foods".

Example: Declare array

```
string[] foods;
```

An array must be **instantiated** to use it. Instantiate means to "create an instance of"; this is done using the "new" keyword. In the example below, the number 5 specifies how many items (often referred to as **elements**) the array can hold.

Example: Instantiate array

```
string[] foods;
foods = new string[5];
```

Example: Instantiate and declare an array at same time

```
string[] foods = new string[5];
```

There are several ways to *assign* values to an array. It is important to point out that arrays start at value 0 and not 1. Listed in the example below are different ways to assign values to an array.

Example: Assign values to an array

```
string[] foods = new string[5];
foods[0] = "Pizza"; // Arrays start at value 0
foods[1] = "Burger";

string[] drinks = new string[5] { "Pepsi", "Sprite", "", "", "" };
// The additional "" are holding the place for the remaining 3 items

string[] cars = new string[] { "Ford", "Toyota" };

string[] animals = { "Cat", "Dog" };
```

Example: Output values from an array

```
string[] animals = { "Cat", "Dog" };
Console.WriteLine(animals[0]); // Output: Cat
Console.WriteLine(animals[1]); // Output: Dog
Console.WriteLine($"The animals are {animals[0]} and {animals[1]}.");
                        // Output: The animals are Cat and Dog.
```

Full Example: Display all items of an array using a foreach loop

```
string[] foods = { "Pizza", "Burger" };

foreach (string food in foods)
{
   Console.WriteLine($"The food item is {food}.");
}
 // Output: The food item is Pizza.
 // Output: The food item is Burger.
```

Full Example: Sort and array length. The **length** property in this example is used to determine how many loops should occur.

```
int[] someNumbers = { 21, 22, 50, 3, 6 };

Array.Sort(someNumbers); // Sort the numbers in ascending order

// someNumbers.Length is the length of the array, in this case 5
for (int i = 0; i < someNumbers.Length; i++)
{
   Console.Write($"{ someNumbers[i]} "); // Output: 3 6 21 22 50
}
```

The above example uses "Array.Sort()". Sort is a method within the Array class that organizes the contents of an array numerically and alphabetically. The Array class has other methods as well, such as "Array.Reverse()".

ACTIVITY: ARRAYS

Activity: In this activity an array stores months that will later be retrieved. Create a Console Application for a Months App that meets the four requirements listed below.

1. Display the following: "Enter a month (1-12) to display the month name or type Display All".

2. If a number from 1-12 is entered, display the month. Example: 2 is February

3. If "Display All" is entered, display all the months from the array.

4. If incorrect information is entered, such as 15, display a message saying "Not a valid input".

The finished output of the program should look like the examples below:

```
C:\Program Files\dotnet\dotnet.exe
Enter a month (1-12) to display the month name or type Display All: 2
2 is February
```

```
Enter a month (1-12) to display the month name or type Display All: Display All
January
February
March
April
May
June
July
August
September
October
November
December
```

```
C:\Program Files\dotnet\dotnet.exe
Enter a month (1-12) to display the month name or type Display All: 15
Not a valid input
```

Need help with the solution? https://www.unQbd.com/Solutions/CSharp5th/Arrays

ARRAYS: 2-D AND 3-D

In the previous section, the arrays have all been **one-dimensional (1-D) arrays**. It is possible to make **two-dimensional (2-d) arrays** , **three-dimensional (3-D) arrays**, and so on. These **Multidimensional arrays** can be useful; the example below shows a two-dimensional array for a game map. Two values are needed for a 2-D array (row and column). Indicate a 2-D array by entering a *comma* between the square brackets.

Example: Declare a2-D array

```
string [,] myGameMap = new string [,]
{
  // Empty 2D array
};
```

Full Example: 2-D array

```
// 2 -D Array for a game map
// Place a comma between the brackets[,] to use a 2-D Array
string [,] myGameMap = new string [,]
{
  {"Scary Room1", "Safe Room2", "Safe Room3'},         //  Row 0
  {"Dangerous Room4", "Safe Room5', "Safe Room6'}, // Row 1
  {"Safe Room7', "Scary Room8', "Safe Room9'}         // Row 2
};

// The first number is the row, the second number is the column
// Keep in mind that array values start at 0
Console.WriteLine(myGameMap[0, 0]); //  Output: Scary Room1
Console.WriteLine(myGameMap[1, 0]); //  Output: Dangerous Room4
Console.WriteLine(myGameMap[2, 2]); //  Output: Safe Room9
```

While a 2-D array can be thought of as making a square with width and height (row and column), a 3-D array can be thought of as making acube to include depth (row, column , and depth). When creating a 3-D array, add an *additional comma* between the square bracketsto indicate the 3-D array is being created

Example: Declare a 3-D array

```
string [,,] arrayExample3D = new string [,,]
{
  // Empty 3D array
};
```

Full Example: 3-D array

```
// Add 2 commas between the square brackets to indicate a 3D array
string[,,] arrayExample3D = new string[,,]
{
  {
    {"This is ", "a 3-D "},        // Row 0 , Column 0, Depth is 0 or 1
    {"array ", "example."}         // Row 0 , Column 1, Depth is 0 or 1
  },
  {
    {"They can ", "become "},      // Row 1 , Column 0, Depth is 0 or 1
    {"complex ", "very quickly."}  // Row 1 , Column 1, Depth is 0 or 1
  },
};

                            // [Row, Column, Depth]
Console.Write(arrayExample3D[0, 0, 0]);  // Output: This is
Console.Write(arrayExample3D[0, 0, 1]);  // Output: a 3 -D
Console.Write (arrayExample3D[0, 1, 0]); // Output: array
Console.Write(arrayExample3D[0, 1, 1]);  // Output: example.
Console.WriteLine();   // Add a space
Console.Write(arrayExample3D[1, 0, 0]);  // Output: They can
Console.Write(arrayExample3D[1, 0, 1]);  // Output: beco me
Console.Write(arrayExample3D[1, 1, 0]);  // Output: complex
Console.Write(arrayExample3D[1, 1, 1]);  // Output: very quickly

// Output:
// This is a 3 -D array example
// The can become complex very quickly
```

Activity: This activity focuses on using a 2-D array for checking the occupancy of hotel rooms. Create a Console Application for a Hotel App that meets the three requirements listed below.

1. Display the following "Which floor do you want to check for occupancy? (0 or 1)".

2. Display the following "Which room do you want to check for occupancy? (0, 1, or 2)".

3. Display the floor, room checked, and who is in the room. If the room is empty, display an Empty message.

The finished output of the program should look like the examples below:

```
C:\Program Files\dotnet\dotnet.exe
Which floor do you want to check for occupancy? (0 or 1): 0
Which room do you want to check for occupancy? (0, 1 or 2): 1
Floor 0 Room 1 occupancy: Zach T.
```

```
C:\Program Files\dotnet\dotnet.exe
Which floor do you want to check for occupancy? (0 or 1): 1
Which room do you want to check for occupancy? (0, 1 or 2): 2
Floor 1 Room 2 occupancy: Empty
```

Need help with the solution? https://www.unQbd.com/Solutions/CSharp5th/TwoDArrays

ARRAYS: JAGGED

A **jagged array** is when arrays of various sizes are put inside another array.

Example: Declare jagged array

```
string[][] jaggedEx = new string[3][];
```

Example: Assign arrays to an array

```
jaggedEx[0] = new string[2]; // 2 element array
jaggedEx[1] = new string[5]; // 5 element array
jaggedEx[2] = new string[8]; // 8 element array
```

Below is a visual representation of the jagged array created in the previous examples. The jagged array contains three arrays of different sizes.

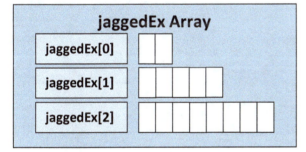

Example: Declare jagged array and assign arrays

```
string[][] jaggedEx = new string[3][] { new string[2], new string[5], new string[8] };
```

Example: Assign value to array

```
jaggedExample[1][4] = "Store this string in Array 1, Element 4";
```

Full Example: Jagged array

```
string[][] classStudents = new string[2][] { new string[2], new string[1] };

classStudents[0][0] = "Sergio";     // In classroom student
classStudents[0][1] = "Jonathan"; // In classroom student
classStudents[1][0] = "Aria";        // Online student

Console.WriteLine($"{classStudents[0][0]} and {classStudents[0][1]}: Classroom students");
// Output: Sergio and Jonathan: Classroom students

Console.WriteLine($"{classStudents[1][0]}: Online student");
// Output: Aria: Online student
```

METHODS

A **method** is a small group of *reusable* code that carries out a single task. To use a method, define the method and **call** (invoke) the method when needed. Methods can be called an unlimited number of times.

The following is the full method syntax of defined method:

public static int Add2Numbers(int num1, int num2)

public Access Modifier	static Static Modifier	int Return Type	Add2Numbers Name	(	int num1 Parameter	,	int num2 Parameter	)

- **Access Modifier :** The access modifier determines the visibility of the method (which classes have access to them). Examples include **public**, **internal** , **private**, and **protected**. If an access modifier is not declared, it is private by default.
- **Static Modifier : Static** methods are shared by all instances of the class, meaning the method can be accessed directly from the class name (static and classes are covered in the "Classes" section). If static is not declared, it is non-static by default.
- **Return Type:** A method can return a value type (Int, String, Bool, etc.) or return **void** if the method does not return anything when called. A return type must be specified.
- **Name:** Method names are required.
- **Parameters:** Arguments (values) are passed to parameters in methods. Parameters are optional.

Example: Method that *returns no values*

```
static void WelcomeMessage()
{
  Console.WriteLine("Hello, Guest");
}

WelcomeMessage(); // Call the method
                  // Output: Hello, Guest
```

Methods can also have **parameters.** Parameters pass values from **arguments** into methods. The parameter refers to the *name* and the arguments are the *values* sent to the method. The example below notes where each is located.

Example: Method that *returns no values* and *accepts a parameter*

```
static void WelcomeMessage(string name) //  Parameter is "name"
{
   Console.WriteLine( $"Hello, {name}");    // Output: Hello, Adam
}

WelcomeMessage("Adam");                  // Call the method and pass the argument "Adam"
```

Parameters can be optional by giving them a default value. In the example below, following the parameter "name", the default value "Guest" is assigned to the parameter making it optional.

Full Example: Method with an optional parameter

```
static void WelcomeMessage(string name = "Guest")
{
   Console.WriteLine( $"Hello, {name}");
}

WelcomeMessage();          // Output: Hello, Guest
WelcomeMessage("Adam"); // Output: Hello, Adam
```

Example: Method that *returns a value* and *accepts a parameter*

```
static int AddThreeNumbers(int num1, int num2, int num3)
{
   int result = num1 + num2 + num3; // Add 3 parameters together
   return result;                   // Return the value
}

int a = 5, b = 2; //  Tip: Data types of the same type can be declared on the same line

int combinedValue = AddThreeNumbers(a, b, 8);
Console.WriteLine( $"{a} + {b} + 8 = {combinedValue}");
// Output: 5 + 2 + 8 = 15

//  The code below demonstrates a method call within WriteLine.
Console.WriteLine( $"{a} + {b} + 6 = {AddThreeNumbers(a, b, 6)}");
// Output: 5 + 2 + 6 = 13
```

Full Example: Passing an *object* to a method (objects and classes covered in the upcoming "Classes" section)

```
Player player1 = new() { Name = "ABC", Health = 100 };

Combat.FightBattle(player1);
Console.WriteLine($"{player1.Name} has {player1.Health} health"); // Output: ABC has 80 health

public class Player
{
  public string Name { get; set; }
  public int Health { get; set; } = 100;
}

public static class Combat
{
  public static void FightBattle(Player thePlayer)
  {
    thePlayer.Health = thePlayer.Health - 20; // Decrease health by 20
  }
}
```

Arguments can be passed by *reference*, not by value. Pass references with the **ref** keyword; changes made in a method for a reference will affect the original value outside the method.

Full Example: Passing a *reference* to a method

```
static void PlayerDamaged(ref int playerHealth)
{
  playerHealth = playerHealth - 20;
}

int playerHealth = 100;
PlayerDamaged(ref playerHealth); // Pass playerHealth as reference parameter.
Console.WriteLine(playerHealth); // Output: 80
```

ACTIVITY: METHODS

Activity: This activity focuses on creating a method to determine if a number is positive or negative. Create a Console Application for a PositiveOrNegative App that meets the three requirements listed below.

1. Display the following "Positive or Negative?" and ask the user to "Enter a number:".

2. Create a method that can determine if a number is positive or negative.

3. Display a message saying the number and if it is *positive*, *negative*, or *neither positive nor negative* (the number 0).

The finished output of the program should look like the examples below:

```
C:\Program Files\dotnet\dotnet.exe
Positive or Negative?
Enter a number: 121
The number 121 is positive
```

```
C:\Program Files\dotnet\dotnet.exe
Positive or Negative?
Enter a number: -5
The number -5 is negative
```

```
C:\Program Files\dotnet\dotnet.exe
Positive or Negative?
Enter a number: 0
The number 0 is neither positive nor negative
```

Need help with the solution? https://www.unQbd.com/Solutions/ CSharp5th/ Methods

CLASSES

A **class** is a group of related methods and variables with common attributes. A class can be thought of as a blueprint or template and **objects** are **instantiated** (created) from it. A single class can instantiate an infinite number of objects.

Example: Create a class

```
class Card
{
  // Note: This example uses get-set properties. The upcoming section
  //       "Get-Set Properties" covers why and how they are used.
  public string Name { get; set; }
  public int Value { get; set; }
  public string Suit { get; set; } = "Heart"; // Set a default value of "Heart"
}
```

Example: Instantiate an object

```
Card Card1 = new();
```

Example: Assign values to an object

```
Card1.Name = "King";
Card1.Value = 13;
```

Example: Instantiate an object and assign values

```
Card Card2 = new()
{
  Name = "Nine",
  Value = 9,
  Suit = "Spade"
};
```

Example: Display output from an object

```
Console.WriteLine($"Card1 Suit: {Card1.Suit} Name: {Card1.Name} Value: {Card1.Value}");
Console.WriteLine($"Card2 Suit: {Card2.Suit} Name: {Card2.Name} Value: {Card2.Value}");
```

Full Example: Class

```
// Example 1: Instantiate an object
Card Card1 = new();

// Assign values to the object
Card1.Name = "King";
Card1.Value = 13;

// Example 2: Instantiate an object and assign values
Card Card2 = new()
{
  Name = "Nine",
  Value = 9,
  Suit = "Spade"
};

Card1.DisplayCard(); // Output: Card Suit: Heart Name: King Value: 13
Card2.DisplayCard(); // Output: Card Suit: Spade Name: Nine Value: 9

// The class named "Card" has attributes for a playing card
class Card
{
  public string Name { get; set; }
  public int Value { get; set; }
  public string Suit { get; set; } = "Heart"; // Set a default value of "Heart"

  public void DisplayCard()
  {
    Console.WriteLine($"Card Suit: {Suit} Name: {Name} Value: {Value}");
  }
}
```

In the example above, the Card class contains the method "DisplayCard()". In the previous examples, all methods have started with "public static". This method leaves out the *static* keyword. Static methods are accessed through the class name directly. In this example, the method is accessed through instances of the class instantiated called "Card1" and "Card2".

Example: Static method call vs instance method call

```
Card.DisplayCard();   //  A static method call
Card1.DisplayCard(); //  An instance method call
```

Example: An array using a class as the type

```
// DeckOfCards Array holds objects of type Card (Card class created in previous examples)
Card[] DeckOfCards = new Card[]
{
  // Create objects of each card needed for a deck of cards
  new Card {Name = "Ace", Value = 1, Suit = "Heart"},
  new Card {Name = "Two", Value = 2, Suit = "Heart"},
  new Card {Name = "Three", Value = 3, Suit = "Heart"}
  // Continue adding cards...
};
```

Activity: This activity focuses on creating a class for a house inventory program. Create a Console Application for a House Inventory App that meets the four requirements listed below.

1. Display the following: "Press Enter to see all available house inventory".

2. Create a class that has the properties: Address, Size, and Price.

3. Create an array to store at least three fictional houses.

4. Create a method to display all of the houses from the array.

The finished output of the program should look like the example below:

```
C:\Program Files\dotnet\dotnet.exe

Press Enter to see all available house inventory

Address: 1234 Allison St. / Size: 1500 / Price: $200000
Address: 4567 Mark St. / Size: 3500 / Price: $450000
Address: 891011 Karen St. / Size: 2500 / Price: $270000
```

Need help with the solution? https://www.unQbd.com/Solutions/CSharp5th/Classes

GET-SET PROPERTIES

Fields within classes should in general be *private* and only allow access when needed. This is where **Get-Set properties** accessors *Get* or *Set* are used to access or change data. Get-Set properties combine aspects of both fields and methods.

Example: Full way to declare a property

```
private string myProperty;  // Declare a private field
public string MyProperty    // Property is Public: it can be accessed from outside the class
{
  get { return myProperty; }  // Get retrieves the value from myProperty and returns it
  set { myProperty = value; } // Set gives access to change private field myProperty
}
```

The "MyProperty" *get* section can return the private field value of "myProperty" and the *set* section can assign a value to the private field "myProperty". It is common to name the private field the same name as the property except with a lowercase letter or with an underscore and then a lowercase letter, for example "_myProperty".

Full Example: Get-Set Property

```
Vehicle brand = new();

brand.Car = "Toyota";            // Use s set in Car property to set private field car to "Toyota"
Console.WriteLine(brand.Car);    // Uses get in Car property to get value of private field car
                                 // Output: Toyota

public class Vehicle
{
  private string car;
  public string Car
  {
    get { return car; }
    set { car = value; }
  }
}
```

It is possible to add additional logic to a property. In the example below, logic within the *get* verifies a number between 1-12 is entered, otherwise 0 is returned.

Full Example: Get-Set property with logic

```csharp
Month TheMonth = new();

TheMonth.TheMonthVerify = 5;
Console.WriteLine(TheMonth.TheMonthVerify); // Output: 5

TheMonth.TheMonthVerify = 13;
Console.WriteLine(TheMonth.TheMonthVerify); // Output: 0

public class Month
{
  private int theMonth;

  public int TheMonthVerify
  {
    get
    {
      // If a number is greater than 12 or less than 0 the value is set to zero.
      if ((theMonth > 12) || (theMonth < 0))
      {
        theMonth = 0;
      }
      return theMonth;
    }
    set
    {
      theMonth = value;
    }
  }
}
```

There is a shorthand way, referred to as **auto-implemented property** , to create get-set properties when no extra logic is needed. The syntax is a lot simpler, and it is no longer necessary to create a private field as this is created behind the scenes.

Example: auto-implemented property

```csharp
public string MyProperty { get; set; }
```

Example: Assign a default value when declaring an auto-implemented property

```csharp
public string MyProperty { get; set; } = 5;
```

Full Example: Auto-implement property

```
Vehicle brand = new();

brand.Car = "Toyota";          // Uses set in Car property to set private field car to "Toyota"
Console.WriteLine(brand.Car);  // Uses get in Car property to get value of private field car
                               // Output: Toyota

public class Vehicle
{
    public string Car { get; set; }
}
```

Using get-set properties gives the ability to define how accessible the property is. It may be tempting to give all properties getters and setters, however in a lot of cases only one is necessary. Always give as little access as possible.

Example: Get only property (read only)

```
public string MyProperty { get; }
```

Leaving out "set" means information can be still be retrieved, but the value cannot be changed.

Another option is to change the accessibility of the set property. In the example below, set has the private modifier before it, meaning it can only be changed within the class.

Example: Private set

```
public int CarID { get; private set; }
```

Full Example: Private set

```
Vehicle car1 = new();
Console.WriteLine(car1.CarID); // Output: 54

public class Vehicle
{
    public void VehicleID(int theID)
    {
        CarID = theID;
    }

    public int CarID { get; private set; }
}
```

When instantiating an object that contains properties you can assign values to them one at a time or can assign multiple at the same time using **object initializer syntax** .

Example: Object initializer syntax

```
Vehicle car1 = new() { Brand = "Toyota", Color = "Red" };
```

Full Example: Object initializer syntax

```
// Example: Object Initializer Syntax
Vehicle car1 = new() { Brand = "Toyota", Color = "Red" };

// Assign 1 at a time
Vehicle car2 = new();
car2.Brand = "Ford";
car2.Color = "White";

public class Vehicle
{
    public string Brand { get; set; }
    public string Color { get; set; }
}
```

The keyword *static* in **static classes**denotes that the class is singular; you cannot create an instance (object) of the class. In a static class all methods and fields must be static.

Example: Static class

```csharp
static class OnlyOneHouse
{
  public static string color { get; set; } = "Red";
}
```

Full Example: Static class

```csharp
OnlyOneHouse.Height = 10;
OnlyOneHouse.Size = OnlyOneHouse.Height * OnlyOneHouse.Width;

OnlyOneHouse.DisplayHouse(); // Output: Size of House: 50 and color Red

static class OnlyOneHouse
{
 // Default attributes for House class
 public static int Height { get; set; }
 public static int Width { get; set; } = 5;
 public static int Size { get; set; } = 25;
 public static string Color { get; set; } = "Red";

 public static void DisplayHouse()
 {
   Console.WriteLine($"Size of House: {Size} and color {Color}");
 }
}
```

CLASSES: CONSTRUCTOR

A **constructor** is essentially a method in a class that executes when the object is instantiated. The constructor will have the same name as the class and can be overloaded. A constructor cannot have a *return type* or be *static*.

Example: Constructor

```csharp
public class Person
{
  public Person()
  {
    // Empty constructor
  }
}
```

Full Example: Constructor

```csharp
Person person1 = new(); // Call constructor with no arguments
Console.WriteLine(person1.Name); // Output: Empty

Person person2 = new("Adam"); // Call constructor with one argument
Console.WriteLine(person2.Name); // Output: Adam

public class Person
{
  public string Name { get; set; }

  public Person() // Constructor with no parameters
  {
    Name = "Empty";
  }

  public Person(string firstName) // Constructor with one parameter
  {
    Name = firstName;
  }
}
```

MINI QUIZ: CLASS CONSTRUCTOR

Mini Quiz

1. A constructor can be in a static class.

    A. True
    B. False

2. A constructor can have a return value.

    A. True
    B. False

3. A constructor has to bethe exact same name as the class name.

    A. True
    B. False

4. A constructor can be overloaded.

    A. True
    B. False

Mini Quiz Answers

1. B
2. B
3. A
4. A

CLASSES: INHERITANCE

Classes can inherit from other classes, meaning they acquire all of another class's features. With class **inheritance,** the **derived** class acquires all the features from a **base** class. C# only supports single inheritance, meaning it can only inherit one class at a time.

Example: Inheritance

```
public class ASecondClass : TheFirstClass
{
  // ASecondClass inherits all the features from the base class TheFirstClass
}
```

Full Example: Inheritance

```
CarSedanDerivedClass Car1 = new();
Car1.Brand = "Tesla";

// An object of CarSedanDerivedClass has inherited information from CarBaseClass
Console.WriteLine( $"A {Car1.Brand} has {Car1.Wheels} wheels and {Car1.Doors} doors");
// Output: A Tesla has 4 wheels and 2 doors

public class CarBaseClass
{
  public string Brand { get; set; }
  public int Wheels { get; } = 4;
}

public class CarSedanDerivedClass : CarBaseClass
{
  public int Doors { get; } = 2;
}
```

C# only supports single inheritance; however, inheritance is transient, meaning it can form a class hierarchy.

Full Example: Inheritance transient

```csharp
BClassExample obj1 = new();
// BClassExample still has access to BaseClassExample because inheritance is transient

Console.WriteLine(obj1.TheBaseProperty); // Output: 5

public class BaseClassExample
{
    public int TheBaseProperty { get; } = 5;
}

public class AClassExample : BaseClassExample
{
    // empty class
}

public class BClassExample : AClassExample
{
    // empty class
}
```

Sealed classes prevent a class from being inherited. In some scenarios using a sealed class will also provide a performance boost.

```csharp
sealed class ExampleBaseClass
{
    // This class cannot be inherited from
}
```

MINI QUIZ: CLASS INHERITANCE

Mini Quiz

1. C# allows for multiple inheritance .

    A. True
    B. False

2. The following code is valid.

```csharp
public class SecondClass: BaseClass, FirstClass
{
  // Empty class
}
```

    A. True
    B. False

3. The following code is valid.

```csharp
public class BaseClass
{
  // Empty class
}

public class FirstClass: BaseClass
{
  // E mpty class
}

public class SecondClass: FirstClass
{
  // E mpty class
}
```

    A. True
    B. False

Mini Quiz Answers

3. A
2. B
1. B

METHODS: OVERLOAD

Overloading a method occurs when methods have the same name but different *signatures*. The compiler is smart enough to know which method to use based on what is being called.

Example: Overload Method

```
class ExampleOverloadMethod
{
  public static void DisplayMe()
  {
    Console.WriteLine( "Result: Nothing to Display" );
  }

  public static void DisplayMe(string theString)
  {
    Console.WriteLine( $"The string {theString} was called');
  }
}
```

Full Example: Overload Method

```
ExampleOverloadMethod.DisplayMe();                  // Output: "Result: Nothing to Display"
ExampleOverloadMethod.DisplayMe("Hello World" );    // Output: "The string Hello World was called"
ExampleOverloadMethod.DisplayMe(215);               // Output: "The int 215 was called"
ExampleOverloadMethod.DisplayMe(215, 500);          // Output: "The total of those 2 numbers is 715"

static class ExampleOverloadMethod
{
  public static void DisplayMe()
  {
    Console.WriteLine( "Result: Nothing to Display" );
  }

  public static void DisplayMe(string theString)
  {
    Console.WriteLine( $"The string {theString} was called');
  }

  public static void DisplayMe(int theInt)
  {
    Console.WriteLine( $"The int {theInt} was called');
  }

  public static void DisplayMe(int theInt1, int theInt2)
  {
    int tempInt = theInt1 + theInt2;
    Console.WriteLine( $"The total of those 2 numbers is {tempInt} ");
  }
}
```

MINI QUIZ: METHOD OVERLOADING

Mini Quiz

1. The following is a valid method overload.

```
class QuizDisplayMe
{
  public static void DisplayMe(string theString)
  {
    // ... method body
  }

  public static void DisplayMe(string theStringTwo)
  {
    // ... method body
  }
}
```

    A. True
    B. False

2. The following is a valid method overload.

```
class QuizDisplayMe
{
  public static void DisplayMe(string theString)
  {
    // ... method body
  }

  public static void DisplayMe(int theStringTwo)
  {
    // ... method body
  }
}
```

    A. True
    B. False

Mini Quiz Answers

1. B – The method signature must be different. Changing the name of a parameter does not change the signature.

2. A

METHODS: RECURSION

The concept of **Recursion** is when something calls itself. When a method calls itself, it is referred to as a **recursive** method.

Example: Recursive method

```
public static double AMethod()
{
                  // Method calls itself
  return AMethod(); // Note: This example would crash a program because it would never end
}
```

Full Example: Recursive method

```
class ExampleRecursive
{
  public static string VerifyWord()
  {
    Console.WriteLine("Choose: Up, Down");
    string direction = Console.ReadLine();

    if (direction == "Up" || direction == "Down")
    {
      return direction;
    }
    return VerifyWord(); // Method calls itself
  }

  string directionToGo = VerifyWord();
}
```

Full Example: Recursive method - Advanced

```
int numberEntered = 4;
int factorial = ExampleRecursive.Factorial(numberEntered);
Console.WriteLine($"Factorial of {numberEntered} = {factorial}");

class ExampleRecursive
{
  // A factorial is the multiplication of every number below it.
  // For example: The factorial of 4 equals (1 * 2 * 3 * 4 = 24)
  public static int Factorial(int number)
  {
    if (number == 0)
    {
      return 1;
    }
    return number * Factorial(number - 1); // Method calls itself
  }
}
```

When a method invokes another method and the resulting method eventually invokes the original method, this is called **indirect** recursion.

Full Example: Indirect recursion

```csharp
int number = 7;
bool theAnswer = IndirectRecursionExample.IsOdd(number);
Console.WriteLine( $"Is the number {number} odd? {theAnswer}");

class IndirectRecursionExample
{
  public static bool IsOdd(int numberSelected)
  {
    if (numberSelected == 0)
    {
      return false;
    }
    return IsEven(Math.Abs(numberSelected)- 1);
  }

  public static bool IsEven(int numberSelected)
  {
    if (numberSelected == 0)
    {
      return true;
    }
    return IsOdd(Math.Abs(numberSelected)- 1);
  }
}
```

METHODS: NAMED ARGUMENTS

Using **named arguments** helps clarify the parameters used when calling a method. Without named arguments, a long list of numbers can quickly become confusing.

Example: Calling a method with a named argument

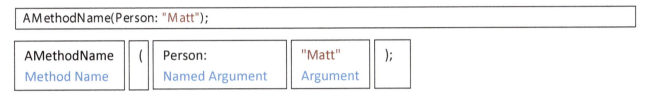

```
AMethodName(Person: "Matt");
```

AMethodName	(	Person:	"Matt"	);
Method Name		Named Argument	Argument	

Full Example: Named arguments

```csharp
// Without named arguments, it is not clear what the numbers represent.
CharacterDescription("Bob", 42, 72, 61);
CharacterDescription("Bob", 42, 72, level: 61);                 // With 1 named argument
CharacterDescription("Bob", age: 42, height: 72, level: 61);    // With 3 named arguments
CharacterDescription(name: "Bob", age: 42, height: 72, level: 61); // With all named arguments

 // TIP: Method arguments can be entered on multiple lines for additional clarity
 CharacterDescription(name: "Bob",
                age: 42,
                height: 72,
                level: 61);

static void CharacterDescription(string name, int age, int height, int level)
{
   Console.WriteLine($"{name}: level {level}, {age} years old, and height of {height} inches.");
}
```

METHOD: VIRTUAL

To *redefine* a method in a derived class, use a **virtual method** . Virtual methods are redefined using the "virtual" keyword in the base class and the "override" keyword in the derived class.

Example: Virtual method

```csharp
public virtual void Sound()
{
    Console.WriteLine("Generic Noise");
}
```

Full Example: Virtual method

```csharp
Animal anyAnimal = new();
anyAnimal.Sound(); // Output: Generic Noise

Dog aDog = new();
aDog.Sound(); // Output: Bark

class Animal
{
    public virtual void Sound()
    {
        Console.WriteLine("Generic Noise");
    }
}

class Dog : Animal
{
    public override void Sound()
    {
        Console.WriteLine("Bark");
    }
}
```

CLASSES: ABSTRACT

Abstract classes are commonly used for *inheritance* and cannot be instantiated. **Abstract methods** within an abstract class must be contained in the child class or the program will not compile. Virtual methods are not required in child classes.

Example: Abstract class

```
public abstract class Talk
{
 // Empty abstract class
}
```

Full Example: Abstract class

```
SpanishLanguage languageSelected = new();
languageSelected.Greeting(); // Output: Hola
languageSelected.Goodbye(); // Output: Goodbye!

public abstract class Talk
{
  // Abstract must be implemented in child class
  public abstract void Greeting();

  // Virtual is NOT required to be in child class but may be optionally overridden.
  public virtual void Goodbye()
  {
    Console.WriteLine("Goodbye!");
  }
}

public class SpanishLanguage : Talk
{
  public override void Greeting()
  {
    Console.WriteLine("Hola");
  }
}
```

CLASSES: PARTIAL

Partial classescan be split across two or more source files. Each file source will contain a section of the class and all parts will be combined when compiled. The "partial" keyword is used for classes, structs, and interfaces, and means that other parts may be defined within the same namespace.

Example: Partial class

```
public partial class PartialExample
{
  // Empty partial class
}
```

Full Example: Partial class

```
Employee Employee1 = new();
Employee1.Name = "John";
Employee1.WelcomeMessage(Employee1.Name);
// Output: John, welcome to our employee information center

public partial class Employee
{
  public string Name { get; set; }
}

public partial class Employee
{
  public void WelcomeMessage(string name)
  {
    Console.WriteLine($"{name}, welcome to our employee information center");
  }
}
```

While you can split up larger classes into partial classes, in most cases creating new classes with more specific sections is advised. Partial classes are generally used for situations that involve "owner" boundaries, such as with a GUI (Graphical User Interface). The GUI has a framework associated with it with it and using partial classes will ensure that the framework does not overwrite any of the code.

ACCESS MODIFIER: PUBLIC, INTERNAL, PROTECTED, AND PRIVATE

Access modifiers control how information can be accessed.

- **Public:** Can be accessed by any other code in the same assembly or another assembly that references it.
- **Internal :** Can be accessed by any other code in the same assembly. Internal is the default *class* modifier if it is not specified.
- **Protected:** Can be accessed only by the code in the same class or in a class that is derived from that class.
- **Private:** Can be accessed only by code in the same class. Private is the default *method* modifier if it is not specified.

Full Example: Public, internal, protected, and private

```
VehicleOption car = new();
car.a = 4; // Able to access data
car.b = 5; // Error: b can only be accessed by classes derived from VehicleOption or within the class.
car.c = 6; // Error: c can only be accessed from within the class

AdvancedOption truck = new();
truck.a = 7; // Able to access data
truck.c = 9; // Error because private can only be accessed from within the class

class AdvancedOption: VehicleOption
{

}

internal class VehicleOption
{
  public int a = 1;
  protected int b = 2;
  private int c = 3;
}
```

MINI QUIZ: ACCESS MODIFIERS

Mini Quiz

1. Which of the following gives the most access?

    A. Internal
    B. Private
    C. Public
    D. Protected

2. What is the default modifier for a method?

    A. Internal
    B. Private
    C. Public
    D. Protected

3. What is the default modifier for a class?

    A. Internal
    B. Private
    C. Public
    D. Protected

4. Which of the following gives the least access?

    A. Internal
    B. Private
    C. Public
    D. Protected

Mini Quiz Answers

1. C
2. B
3. A
4. B

RANDOM NUMBER

It is often necessary to create a random number in software applications. Below are several techniques on how to create a random number using the Random class.

Example: Random number from *0 to 2,147,483,646*

```
Random r = new(); // Instantiate object r from the Random class

// The Next() method within the r object returns a number from 0 to Int32.MaxValue
int myRandomNumber = r.Next();
```

Example: Argument in Next method – "maxValue"

```
Random r = new(); //  Instantiate object r from the Random class

// The Next method returns a non -negative random integer that is less than the
// specified maximum value supplied.
int myRandomNumber = r.Next(11);  // Output: Will vary from 0 to 10
```

Example: Arguments in Next Method - "minValue" and "maxValue"

```
Random r = new(); // Instantiate object r from the Random class

int myRandomNumber = r.Next(10, 21);  // (minValue, maxValue)
                                       // minValue is greater than or equal to.
                                       // maxValue is less than the specified maximum.
                                       // Output: Will vary from 10 to 20
```

Full Example: Random number in a method

```
Console.WriteLine(RandomNumMethod( 5001)); // Output: Will vary from 1 to 5000

static int RandomNumMethod( int num1)
{
   Random r = new(); // Instantiate object r from the Random class
   return  r.Next(1, num1);
}
```

ACTIVITY: RANDOM NUMBER

Activity: This activity focuses on using a random number for a guessing game. Create a Console Application for a Guessing Game App that meets the four requirements listed below.

1. Display the following: "Guess the computer's number between 1 and 10. You have 3 attempts!".

2. Have the computer select a random number from 1-10.

3. Have the ability to guess the computer's number 3 times.

4. Display a message saying if the guess was low/high/correct.

The finished output of the program should look like the examples below:

```
C:\Program Files\dotnet\dotnet.exe
Guess the computer's number between 1 and 10. You have 3 attempts!
Guess number 1: 3
You guessed too low!
Guess number 2: 8
You guessed too high!
Guess number 3: 6
You guessed too high!
You LOSE, the number was 4
```

```
C:\Program Files\dotnet\dotnet.exe
Guess the computer's number between 1 and 10. You have 3 attempts!
Guess number 1: 5
You guessed too low!
Guess number 2: 8
You guessed correctly!
You WIN, the number was 8
```

Need help with the solution? https://www.unQbd.com/Solutions/CSharp5th/RandomNumber

MEMORY: HEAP AND STACK

When a variable is declared, a portion of memory is allocated in **RAM**. This portion of memory contains the *name*, *data type*, and *value* of the variable. Depending on the data type, the portion of memory allocated will be either **stack** or **heap** memory.

The *stack* is used to keep track of local variables and the program's state. The *heap* is used to store data that can be accessed anytime and from anywhere in your program.

Data types are either **value types** or **reference types**. The most common *reference* types are: strings, arrays, and objects which are on the heap; everything else is a value type on the stack.

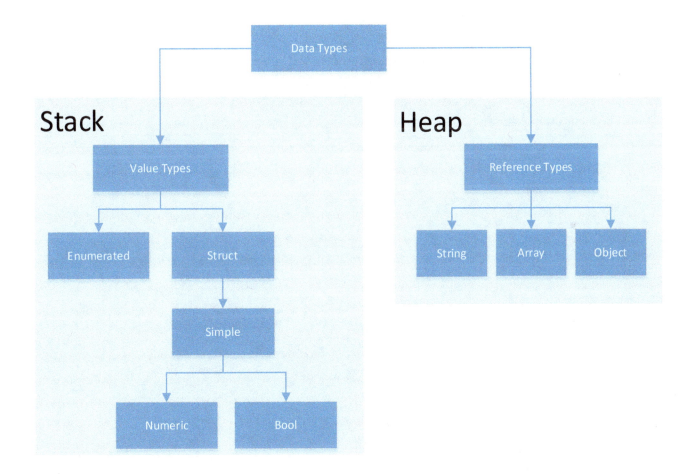

Reference types can be assigned **null reference**, meaning a reference can point to nothing at all. This is accomplished using the "null" keyword. Assigning null to ordinary value types is not allowed.

Full Example: Memory – stack and heap

```csharp
int a = 7;      // Value type, Stack
int b = a;      // Value type, Stack
bool c = true; // Value type, Stack

string d = "Hello";    // Reference type, Heap
int[] e = new int[20]; // Reference type, Heap

// Instantiates an object: Reference type, Heap
MyClass theObject = new();
MyClass theSecondObject = theObject;

// Objects: Reference type, Heap
theObject.g = 5;
theSecondObject.g = 9;

// Changing theSecondObject to 9 changed both objects to 9 because it is a reference.
Console.WriteLine(theObject.g);        // Output: 9
Console.WriteLine(theSecondObject.g); // Output: 9

public class MyClass
{
    public int g { get; set; }
}
```

MINI QUIZ: MEMORY - HEAP AND STACK

Mini Quiz

1. Value types can be assigned a null reference.

    A. True
    B. False

2. An Int is a value type.

    A. True
    B. False

3. A string is a value type.

    A. True
    B. False

4. An object is a reference type.

    A. True
    B. False

Mini Quiz Answers

1. B
2. A
3. B
4. A

MEMORY: BOXING VS UNBOXING

When data moves from *value type* to a *reference type*, it is called **boxing**, and going from *reference type* to *value type* is called **unboxing**.

Example: Boxing

```
int a = 123;  // Value type
object b = a; // Boxing: Value type to a reference type
```

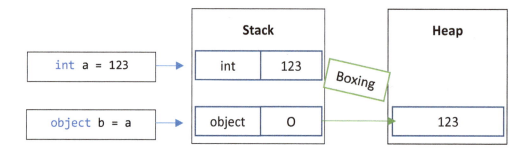

Example: Unboxing

```
int a = 123;   // Value type
object b = a;  // Boxing: Value type to a reference type
int c = (int)b; // Unboxing: Reference type to value type
```

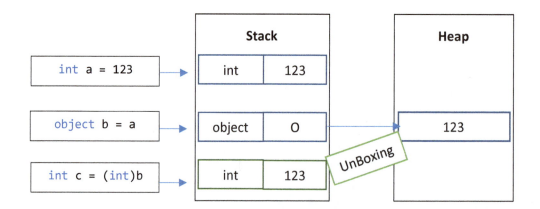

STRUCTS

A **struct** is similar to a class and is useful for *lightweight objects*, such as point, rectangle, and color. Lightweight objects can also be created with classes, however it is often more memory efficient to use a struct. Unlike classes, structs are *value* types and not *reference* types.

Example: Struct

```
public struct TheLocation
{
 // Empty struct
}
```

Full Example: Struct

```
TheLocation theLocation1 = new();
Console.WriteLine($"Location1: x = {theLocation1.X}, y = {theLocation1.Y}");
// Output: Location1: x = 0, y = 0

TheLocation theLocation2 = new(23, 6);
Console.WriteLine($"Location2: x = {theLocation2.X}, y = {theLocation2.Y}");
// Output: Location2: x = 23, y = 6

public struct TheLocation
{
  public int X { get; private set; }
  public int Y { get; private set; }

  public TheLocation(int point1, int point2)
  {
    X = point1;
    Y = point2;
  }
}
```

A struct can be declared without using the keyword *new*. This is unique to structs and this would not work if it were a class. All value types (Int, Char, Bool, etc.) are structs, which is how you can declare them without using the keyword "new".

Full Example: Declaring a struct without using the keyword "new"

```
// Previous way
// TheLocation theLocation2 = new (23, 6);

TheLocation theLocation3;
theLocation3.X = 23;
theLocation3.Y = 6;

Console.WriteLine( $"Location1: x = {theLocation3.X} , y = {theLocation3.Y}");

public struct TheLocation
{
  public int X;
  public int Y;

  public TheLocation(int point1, int point2)
  {
    X = point1;
    Y = point2;
  }
}
```

RECORDS

A **record** is a type that makes it easier to create *immutable* reference types and provides equality comparisons.

Example: Record

```
public record Student(string Name, string Assignment, int Grade);
```

Example: Instantiate a record

```
Student student1 = new("Pablo", "Exam 1", 97);
```

It is important to point out that when records are created in the above manner they are immutable, meaning they can not be changed.

```
student1.Grade = 98; // Error because the value can not be changed
```

The "with" keyword can be used to create a *copy* of a record and change the properties as needed.

Example: With keyword

```
Student student2 = student1 with { Grade = 98 };

// All the original properties of student1 are copied to student2, only Grade is changed
```

Records have equality built in which is *value-based* and includes a check that the types match as well.

Example: Equality

```
Student student1 = new("Pablo", "Exam 1", 97);
Student student2 = new("Pablo", "Exam 1", 97);
Student student3 = new("Jason", "Exam 1", 95);
Teacher teacher1 = new("Pablo", "Exam 1", 97);

bool match1 = student1.Equals(student2); // True
bool match2 = student1.Equals(student3); // False
bool match3 = student1.Equals(teacher1); // False, the types do not match
```

Example: Textual representation

```
Console.WriteLine(student1.ToString());

// Output: Student { Name = Pablo, Assignment = Exam 1, Grade = 97 }
```

Records can also use the **deconstruct** method to separate the record into component properties

Example: Deconstruct

```
var (studentName, studentAssignment, studentGrade) = student1; // Deconstruct
Console.WriteLine(studentName);
Console.WriteLine(studentAssignment);
Console.WriteLine(studentGrade);
```

Example: Record with a method

```
public record Student(string Name, string Assignment, int Grade)
{
  public void DisplayMessage()
  {
    Console.WriteLine("Hello");
  }
}
```

Records can inherit from other records.

Example: Inheritence

```
public record ClassRoom(string Name, string Assignment, int Grade, int Room) : Student(Name,
Assignment, Grade);
```

Full Example: Record

```
Student student1 = new("Pablo", "Exam 1", 97);
Student student2 = new("Jason", "Exam 1", 95);

Console.WriteLine("Are student1 and student2 the same?");
Console.WriteLine(student1. Equals(student2)); // Output: false

ClassRoom student3 = new("Pablo", "Exam 1", 97, 101);
 student3.DisplayMessage(); // Output: Hello

public record Student(string Name, string Assignment, int Grade)
{
   public void DisplayMessage()
   {
     Console.WriteLine("Hello");
   }
}

public record ClassRoom(string Name, string Assignment, int Grade, int Room) : Student(Name,
Assignment, Grade);
```

NAMESPACES & USING DIRECTIVE

A **namespace** helps organize a large group of related code. To use a namespace, call it with the **using directive** or include it before the class name.

Example: Namespace

```
namespace NamespaceExample1Console
{
  // Empty Namespace
}

// In C# 10 file scoped namespaces were introduced. The namespace below does the same as the above
namespace and will include everything in the file.
namespace NamespaceExample2Console;
```

The example below uses the "System" namespace. Within the System namespace there is a class called Console, and within the Console class there is the method WriteLine.

Full Example: Using a Namespace

```
using System; // C# 10 comes with Implicit Usings, meaning the most popular using namespaces such as
System are already built in so there is no need to declare them

Console.WriteLine("Hello World!");

// If "using System;" was not used, the fully qualified name must be used
System.Console.WriteLine("Hello World!");
```

Global usings were introduced in C# 10. When the word global appears before the using directive, that using statement will now apply to the entire project. It is common to create a GlobalUsings.cs file and include all of the global usings in one location.

Example: Global using

```
global using System;
```

The *static* members of a *static* class can be accessed without having to specify the type name, meaning "Console.WriteLine ()" can be simplified to just "WriteLine ()". This is done with the "using static" directive.

Full Example: Using static directive: "System.Console"

```
using static System.Console;

WriteLine("Console.WriteLine() is simplified to just WriteLine()");
ReadLine(); // Console.Readline() is simplified to just ReadLine();
```

Full Example: Namespace created and called

```
Console.WriteLine(Automobiles.Cars.theCar);

namespace Automobiles
{
  static class Cars
  {
    public static string theCar = "ToyotaCar";
  }

  class Trucks
  {
    // Empty class to demonstrate namespaces can contain multiple classes
  }
}
```

In the example above, within WriteLine "Automobiles.Cars.theCar" the full name had to be used. However, if the static class directive is used for "Automobiles.Cars", this can now be shortened to just "theCar" as demonstrated in the example below.

Full Example: Creating and using a static directive

```
using static Automobiles.Cars;

Console.WriteLine(theCar);
Console.ReadLine();

namespace Automobiles
{
  static class Cars
  {
    public static string theCar = "Toyota Car";
  }
}
```

GENERICS

Generics define type-safe classes without committing to any specific data types. They are essentially a placeholder until a specified data type is declared. There are a lot of prebuilt classes using generics. This section covers how to create generic classes. The next section, "List", demonstrates how to use a prebuilt generic class.

The example below demonstrates the use of the **generic type parameter**, which uses the angle brackets "<>". Generic type names are usually a single capital letter, "T" being the most common, or a simple name starting with "T" such as "TKey" or "TValue".

Example: Generic class

```
class AGenericClass<T>
{
   // Empty generic class
}
```

Assume in the next example it is being instantiated with a string in the following way:

```
AGenericClass<string> stringExample = new("Hello");
```

Example: Generic class

```
class AGenericClass<T>
{
  private T aVariable;

  public AGenericClass(T aValue)
  {
    aVariable = aValue;
    Console.WriteLine($"The value is {aVariable} and is a {typeof(T).ToString()}");
    // Output: The value is Hello and is a System.String
  }

  public void DisplayTypeOnly(T aValue)
  {
    Console.WriteLine($"A {typeof(T).ToString()}"); // Output: A System.String
  }
}
```

In the example above, the generic type parameter "T" uses string as the data type, because that is how the class was instantiated. If the class were instantiated with the data type int then all the corresponding parameters would have used int instead.

Full Example: Generics

```csharp
AGenericClass<int> intExample = new(5);
// Output: The value is 5 and is a System.Int32

AGenericClass<string> stringExample = new("Hello");
// Output: The value is Hello and is a System.String

AGenericClass<double> doubleExample = new(5.5);
// Output: The value is 5.5 and is a System.Double

intExample.DisplayTypeOnly(5);            // Output: A System.Int32
stringExample.DisplayTypeOnly("Hello"); // Output: A System.String
doubleExample.DisplayTypeOnly(5.5);      // Output: A System.Double

class AGenericClass<T>
{
  private T aVariable;

  public AGenericClass(T aValue)
  {
    aVariable = aValue;
    Console.WriteLine($"The value is {aVariable} and is a {typeof(T).ToString()}");
  }

  public void DisplayTypeOnly(T aValue)
  {
    Console.WriteLine($"A {typeof(T).ToString()}");
  }
}
```

LIST

Lists and arrays are similar in that they both hold items. Lists can easily search, sort, add, and remove items because of built-in functionalities. Lists can constantly be changed to include more or fewer items, unlike an array that has a set number of items.

To use a list, add the following line of code.

```
using System.Collections.Generic; // C# 10 comes with Implicit Usings, meaning the most popular using namespaces such as this one are already built in so there is no need to declare it
```

The code above gives access to the *Generic* namespace which includes the generic List class. Because the example is using generics specify the data type of the List class such as string, int, or double.

Example: Instantiate an empty list

```
List<string> foods = new();
```

Example: Assign items to a list

```
List<string> foods = new();
foods.Add("Pizza");
foods.Add("Burger");
```

Example: Instantiate and assign items

```
List<string> foods = new() {"Pizza", "Burger"};
```

Example: Insert at a specific index

```
foods.Insert(5, "Cake");
// Pushes back all items after index 5, items in front of 5 stay the same.
```

Example: Remove items from a list

```
foods.Remove("Pizza");
```

Example: Remove at a specific index

```
foods.RemoveAt(5);
```

Example: Output an item

```
Console.WriteLine(foods[0]);
```

Example: Loop through a list

```
foreach (var food in foods)
{
    Console.WriteLine(food);
}
```

Example: Get the size of a list

```
Console.WriteLine(foods.Count);
```

Example: Sort a list alphabetically and numerically

```
foods.Sort();
```

Example: Remove all items from a list

```
foods.Clear();
```

Example: Convert a list to an array

```
string[] foodsArray = foods.ToArray();
```

Full Example: List

```csharp
// Instantiate and assign items to a list
List<string> foods = new()
{
  "Pizza",
  "Burger",
  "Hot Dog"
};

Console.WriteLine(foods[1]); // Output: Burger

// Loop through entire list and output each item
foreach (var food in foods)
{
  Console.WriteLine(food);
  // Output: Pizza
  // Output: Burger
  // Output: Hot Dog
}
```

ACTIVITY: LIST

Activity: This activity focuses on using a list for a grocery list program. Create a Console Application for a Grocery List App that meets the three requirements listed below.

1. Display the following: "Add/Remove/Display items from grocery list (Add/Remove/Display/Exit):".

2. Allow a user to add or remove as many items as they want to a grocery list.

3. Allow a user to display all items on the grocery list.

The finished output of the program should look like the example below:

```
C:\Program Files\dotnet\dotnet.exe
Grocery List

Add/Remove/Display items from grocery list (Add/Remove/Display/Exit): Add
What would you like to add? Chicken
Chicken has been added.

Add/Remove/Display items from grocery list (Add/Remove/Display/Exit): Add
What would you like to add? Fish
Fish has been added.

Add/Remove/Display items from grocery list (Add/Remove/Display/Exit): Display
Grocery List:
Chicken
Fish

Add/Remove/Display items from grocery list (Add/Remove/Display/Exit): Remove
What would you like to remove? Chicken
Chicken has been removed.

Add/Remove/Display items from grocery list (Add/Remove/Display/Exit): Display
Grocery List:
Fish

Add/Remove/Display items from grocery list (Add/Remove/Display/Exit): Exit
```

Need help with the solution? https://www.unQbd.com/Solutions/CSharp5th/List

LINKEDLIST

A **LinkedList** is generally slower than a regular list, however it can be beneficial when adding/removing items in the middle of a list. As with lists, a LinkedList uses generics.

To use a LinkedList, add the following line of code.

```csharp
using System.Collections.Generic; // C# 10 comes with Implicit Usings, meaning the most popular using namespaces such as this one are already built in so there is no need to declare it
```

Example: LinkedList

```csharp
LinkedList<string> ingredientsOrder = new();
```

Full Example: LinkedList and LinkedListNode

```csharp
LinkedList<string> ingredientsOrder = new();

ingredientsOrder.AddLast("Add an egg");
ingredientsOrder.AddLast("Add butter");
ingredientsOrder.AddFirst("Get a bowl");

// Find the positions of "Get a bowl"
LinkedListNode<string> node = ingredientsOrder.Find("Step1: Get a bowl");
ingredientsOrder.AddAfter(node, "Step2: Add sugar");

foreach (string item in ingredientsOrder)
{
  Console.WriteLine(item);
  // Output: Get a bowl
  // Output: Add sugar
  // Output: Add an egg
  // Output: Add butter
}
```

DICTIONARY

A **dictionary** consists of a **key** and a **value**. In an array, the key is created automatically (0, 1, ..); with dictionaries the key name can be set to something meaningful. As with lists, dictionaries use generics.

To use a dictionary, add the following line of code.

```
using System.Collections.Generic; // C# 10 comes with Implicit Usings, meaning the most popular using
namespaces such as this one are already built in so there is no need to declare it
```

Dictionaries have two generic data type parameters. Because they are generic, any data type can be entered. In the example below, the first generic data type is a *string* and is used as the key. The second generic data type is an *int* and is used for the value.

Example: Instantiate a dictionary

```
Dictionary <string, int> myInventory = new();
```

Example: Assign items to a dictionary

```
Dictionary <string, int> myInventory = new();
myInventory.Add( "Pens", 7);
myInventory.Add( "Computers", 2);
```

Example: Instantiate and assign items to a dictionary

```
Dictionary <string, int> myInventory = new()
{
  { "Pens", 7 },
  { "Computers", 2 }
};
```

Example: Remove items from a dictionary

```
myInventory.Remove("Pens");
```

Example: Update an item

```
myInventory["Computers"] = 6;
```

Example: Output an item

```
Console.WriteLine($"Key: {myInventory["Computers"]}");
```

To loop through a dictionary, use a ForEach loop and the keyword "KeyValuePair<TKey, TValue>". This is needed because dictionary elements are retrieved as **KeyValuePair** objects.

Example: Loop through a dictionary

```
foreach (KeyValuePair<string, int> theItem in myInventory)
{
    Console.WriteLine($"{theItem.Key}: quantity {theItem.Value} ");
}
```

Example: Get the size of a dictionary

```
Console.WriteLine(myInventory.Count);
```

Example: Remove all items from a dictionary

```
myInventory.Clear();
```

Full Example: Dictionary

```
Dictionary<string, int> myInventory = new()
{
    { "Pens", 7 },
    { "Computers", 2 }
};

Console.WriteLine(myInventory["Pens"]); // Output: 7

// Loop through entire dictionary and output items
foreach (KeyValuePair<string, int> theItem in myInventory)
{
    Console.WriteLine($"{theItem.Key}: quantity {theItem.Value} ");
}
// Output: Pens: quantity 7
// Output: Computers: quantity 2
```

ENUMERATION

An **enum** (enumeration) is a value type that consists of a collection of **named constants.** Using enums creates a modular design that enhances clarity and reduces the probability of invalid constants. Enums are generally defined directly in the namespace (outside of any class).

Example: Declare an enum

```
enum Direction { North, East, South, West };
```

Example: Assign PlayerDirection an enum value

```
Direction PlayerDirection = Direction.North;
```

Example: Output enum value

```
Direction PlayerDirection = Direction.North;
Console.WriteLine(PlayerDirection); // Output: North
```

Example: Enum value in an If statement

```
if (PlayerDirection == Direction.North)
{
  Console.WriteLine("The player heads north");
}
else if (PlayerDirection == Direction.East)
{
  Console.WriteLine("The player heads east into the woods.");
}
```

Example: Enum in a ForEach loop

```
foreach (Direction theEnumItem in Enum.GetValues(typeof(Direction)))
{
  Console.WriteLine(theEnumItem);
}
```

Full Example: Enum

```csharp
EnumNamespaceExample.EnumExample display = new();
display.Directions();

namespace EnumNamespaceExample
{
  enum Direction { North, East, South, West };

  class EnumExample
  {
    Direction PlayerDirection = Direction.North;

    public void Directions()
    {
      Console.Write("You can go the following directions: ");

      foreach (Direction theEnumItem in Enum.GetValues(typeof(Direction)))
      {
        Console.Write($"{theEnumItem} "); // Output: North East South West
      }

      if (PlayerDirection == Direction.North)
      {
        Console.WriteLine("The player heads north");
      }
      else if (PlayerDirection == Direction.East)
      {
        Console.WriteLine("The player heads east into the woods.");
      }
    }
  }
}
```

By default, enum values start at 0. However, numbers can be assigned to enum values.

Full Example: Assign enum numbers

```csharp
EnumAssignNumbersExample.EnumExample.MonthNumber();

namespace EnumAssignNumbersExample
{
  enum Month { January = 1, February = 2, March = 3, April = 4, May = 5 }; // and so on..

  class EnumExample
  {
    public static void MonthNumber()
    {
      Month value = Month.February;
      int monthNumber = (int)value;
      Console.WriteLine(monthNumber); // Output: 2
    }
  }
}
```

MINI QUIZ: ENUMERATION

Mini Quiz

1. Enums are generally created inside classes.

    A. True
    B. False

2. The following code is valid.

```
enum Size{ Small, Medium, Large, Extra Large, XXL };
```

    A. True
    B. False

3. Enums are often used to enhance clarity and reduce the risk of invalid constants.

    A. True
    B. False

4. Enum values can be assigned like the example below

```
enum Rating { Horrible = 1, OK = 2, Good = 3, Great = 4, Amazing = 5 };
```

    A. True
    B. False

Mini Quiz Answers

1. B
2. B – "Extra Large" must be 1 word with no spacing "ExtraLarge"
3. A
4. A

TERNARY OPERATOR

The **ternary** operator tests a condition by comparing two values and returns a third value based on the results. A ternary operator can often simplify a traditional "If-Else" statement.

Ternary Format:

```
Condition ? First_Result : Second_Result
```

Example: Ternary operator

```csharp
string morningOrNight = (2 < 12) ? "Morning" : "Night"; // Morning is assigned
```

The example below demonstrates how the ternary operator can be used to get the same result with less coding.

Example: "If-Else" statement vs ternary operator

```csharp
int myAge = 23;
string theResponse;

// If -Else statement example
if (myAge < 21)
{
  theResponse = "Under 21 years old";
}
else
{
  theResponse = "Over 21 years old";
}

Console.WriteLine(theResponse); // Output: Over 21 years old

// Ternary operator example
theResponse = (myAge < 21) ? "Under 21 years old" : "Over 21 years old";

Console.WriteLine(theResponse); // Output: Over 21 years old
```

Full Example: Ternary operator in a method

```csharp
Console.WriteLine(AgeCheck(19)); // Output: "Under 21 years old"

static string AgeCheck(int theAge)
{
  return (theAge < 21) ? "Under 21 years old" : "Over 21 years old";
}
```

DATETIME

DateTime is a struct that is used to express an instant in time that includes the date and time of day.

In the example below, an object is created that contains the date and time of the local computer by using "DateTime" and "Now".

Example: Assign current date/time with 'Now'

```
DateTime todaysDate = DateTime.Now;
Console.WriteLine(todaysDate); // Output: 11/10/2022 8:59:22 AM
                               // Will vary based on your current date/time
```

Example: Create a specific date

```
DateTime dob = new DateTime(1984, 1, 20); // (Year, Month, Day)
Console.WriteLine(dob);                    // Output: 1/20/1984 12:00:00 AM
```

Example: Create a specific date and time

```
DateTime dob = new DateTime(1984, 1, 20, 2, 30, 5); // (Year,Month,Day,Hour,Minute,Second)
Console.WriteLine(dob);                              // Output: 1/20/1984 2:30:05 AM
```

There are many ways to format the date and time. A few example are listed below.

Example: Display/format date or time

```
DateTime dob = new(1984, 1, 20, 2, 30, 5);   // (Year,Month,Day,Hour,Minute,Second)
Console.WriteLine($"The Month is: {dob.Month}");      // Output: 1
Console.WriteLine($"The Month is: {dob:MM}");         // Output: 01
Console.WriteLine($"The Month is: {dob:MMM}");        // Output: Jan
Console.WriteLine($"The Month is: {dob:MMMM}");       // Output: January
Console.WriteLine($"The date is: {dob.Day}");         // Output: 20
Console.WriteLine($"The date is: {dob:d}");           // Output: 01/20/1984
Console.WriteLine($"The date is: {dob:D}");           // Output: Friday, January 20, 1984
Console.WriteLine($"The time is: {dob.TimeOfDay}");   // Output: 02:30:05
Console.WriteLine($"The time is: {dob:H:mm}");        // Output: 2:30
Console.WriteLine($"The time is: {dob:H:mm tt}");     // Output: 2:30 AM
```

Once a DateTime object is created, there are ways to add and subtract time from it. In the example below, a month is added using "AddMonths".

Example: Add time

```
DateTime dob = new(1984, 1, 20, 2, 30, 5);              // (Year,Month,Day,Hour,Minute,Second)
Console.WriteLine($"The Month: {dob.Month}");           // Output: The Month: 1

DateTime addMonthToDOB = dob.AddMonths(1);              // Add a month
Console.WriteLine($"The Month: {addMonthToDOB.Month}"); // Output: The Month: 2
```

DateTime objects can be compared to each other to determine if one is earlier, later, or if they are the same.

Full Example: Compare

```
DateTime firstDate = new(2021, 1, 23);
DateTime secondDate = new(2022, 1, 23);

int result = DateTime.Compare(firstDate, secondDate);

if (result < 0)
{
  Console.WriteLine("The first date is earlier");
}
else if (result > 0)
{
  Console.WriteLine("The first date is later");
}
else
{
  Console.WriteLine("The dates are the same");
}

// Dates can also be compared without using the Compare method
if (firstDate < secondDate)
{
  Console.WriteLine("The first date is earlier");
}
```

Activity: This activity focuses on using a DateTime to calculate on what day of the week you were born. Create a Console Application for a Born Day of Week App that meets the three requirements listed below.

1. Ask the questions:

- What year were you born?
- What month were you born?
- What day were you born?

2. Create a DateTime object from the answers.

3. Display the day of week born.

The finished output of the program should look like the example below:

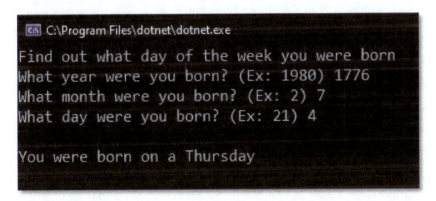

Need help with the solution? https://www.unQbd.com/Solutions/ CSharp5th/ DateTime

THIS (REFERENCE)

The "**this**" keyword refers to the current instance of the class; it helps to qualify members hidden by the same name. The "this" keyword can also be used to pass the entire instance of the current class to a parameter as an object.

Example: this

```csharp
private string name = "Adam";

public void DisplayName(string name = "David")
{
  Console.WriteLine(this.name); // Output: Adam
  Console.WriteLine(name);      // Output: David
}
```

Full Example: this

```csharp
NameClass example1 = new();
example1.DisplayName("David");

class NameClass
{
  private string name = "Adam";
  public string FirstName { get; set; } = "Samantha";
  public string LastName { get; set; } = "West";

  public void DisplayName(string name)
  {
    Console.WriteLine(this.name);                 // Output: Adam
    Console.WriteLine(name);                       // Output: David
    Console.WriteLine(NameHelp.CombineName(this)); // Send entire instance of current class
  }                                                // to another class
}

static class NameHelp
{
  public static string CombineName(NameClass name)
  {
    return $"{name.FirstName} {name.LastName}"; // Output: Samantha West
  }
}
```

METHODS: EXTENSION

Extension methods define a method that appears to belong to a class to which it does not, in fact, belong. Extension methods must be defined in static classes and be static methods.

The built in string class contains methods such as: "ToLower()" and "Length()". In the example below, an extension method called "ToStarBox()" is created and can be used like the previous string methods mentioned. It will appear to be in the string class, when it is not.

Extension method parameters must start with the keyword "this" followed by the type that is being created, in our example the string data type.

Example: Create extension method

```
public static class StringExtensionExample
{
  public static void ToStarBox(this string text)
  {
    string starLine = "**";

    for (int i = 0; i < text.Length; i++)
    {
      starLine += "*";
    }

    Console.WriteLine(starLine);
    Console.WriteLine($"*{text}*");
    Console.WriteLine(starLine);
  }
}
```

Example: Call extension method

```
string aWord = "Hello";

aWord.ToStarBox();
// Output: *******
//         *Hello*
//         *******
```

Full Example: Extension Method

```csharp
string aWord = "Hello";
aWord.ToStarBox();
// Output: *******
//         *Hello*
//         *******

public static class StringExtensionExample
{
  public static void ToStarBox(this string text)
  {
    string starLine = "**";

    for (int i = 0; i < text.Length; i++)
    {
      starLine += "*";
    }

    Console.WriteLine(starLine);
    Console.WriteLine($"*{text}*");
    Console.WriteLine(starLine);
  }
}
```

ACTIVITY: METHODS - EXTENSION

Activity: This activity focuses on using an extension method to find out if a number is lucky. Create a Console Application for Lucky Number App that meets the three requirements listed below.

1. Display the prompt: "Enter a number to find out if it is lucky: ".

2. Create an extension method for an int that that returns the string "Lucky" or "Not Lucky" (in the example below the lucky numbers are 7 and 13).

3. Display if the number is lucky or not lucky.

The finished output of the program should look like the examples below:

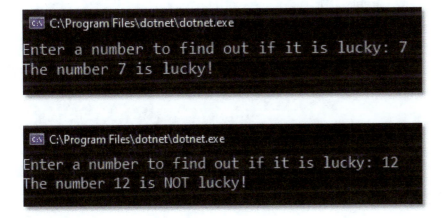

Need help with the solution? https://www.unQbd.com/Solutions/CSharp5th/ExtensionMethod

TUPLES

A **Tuple** can be thought of much like an array, however a Tuple collection can contain different data types. Tuple classes are *generic* containers and can hold between 1 to 8 items.

Example: Instantiate a Tuple

```
Tuple<int, string, int[]> myTuple = Tuple.Create(101, "Hello", new int[] { 41, 52 });
```

To access the information within a Tuple, use the name of the Tuple and then ".Item1", ".Item2" , and so on.

Example: Instantiate and access a Tuple

```
// Instantiate a 3-Item Tuple: Int, String, and an Array of Ints
Tuple<int, string, int[]> myTuple = Tuple.Create(101, "Hello", new int[] { 41, 52 });

// Access a Tuple item
Console.WriteLine(myTuple.Item1);  // Output: 101
Console.WriteLine(myTuple.Item2);  // Output: Hello

foreach (int theNumber in myTuple.Item3)
{
  Console.Write( $"{theNumber} "); // Output 41 52
}
```

Example: Tuple in a method

```
int num1 = 10;
int num2 = 4;

Tuple <int, int> answer = DivideGetQuotientAndRemainder(num1, num2);
Console.WriteLine( $"{num1} / {num2} = {answer.Item1}, with remainder of {answer.Item2}");
// Output: 10 / 4 = 2, with remainder of 2

static Tuple <int, int> DivideGetQuotientAndRemainder( int dividend, int divisor)
{
  int quotient = dividend / divisor;
  int remainder = dividend % divisor; // Uses the modulus operator (%) to get remainder

  return Tuple.Create(quotient, remainder);
}
```

MINI QUIZ: TUPLES

Mini Quiz

1. A Tuple can hold an unlimited number of items.

   A. True
   B. False

2. A Tuple can hold any datatype.

   A. True
   B. False

3. A Tuple can store a List of strings, List of ints, and an Array of strings.

   A. True
   B. False

4. The following code would create a Tuple.

```
Tuple<string, int, int[]> myTuple = Tuple.Create(408, 505, new int[] { 7, 13 });
```

   A. True
   B. False

Mini Quiz Answers

1. B
2. A
3. A
4. A

STRINGBUILDER

When the value of a string is changed, a new string object in memory is created. This is because a string is **immutable** (meaning cannot be changed once created). This uses additional system resources. **StringBuilder** is a **mutable** string-like object whose value will change based on the sequence of characters (a new string object in memory is not created when appended).

To use StringBuilder, include the using statement:

```
using System.Text;
```

Example: Instantiate a StringBuilder

```
StringBuilder theBuilder = new("Hello World!" );
```

Example: Append a StringBuilder

```
StringBuilder theBuilder = new("Hello World!" );
theBuilder.Append(" Today is going to be great!");
Console.WriteLine(theBuilder);  // Output: "Hello World! Today is going to be great!
```

Example: Insert a StringBuilder

```
StringBuilder theBuilder = new("Hello how are you?");
theBuilder.Insert(5, " Jason,");
Console.WriteLine(theBuilder);  // Output: "Hello Jason, how are you?
```

Example: Remove a StringBuilder at a specified index with specified length

```
StringBuilder theBuilder = new("Hello how are you?");
theBuilder.Remove(5, 13);
Console.WriteLine(theBuilder);  // Output: "Hello"
```

Example: Replace part of a StringBuilder

```
StringBuilder theBuilder = new("Hello, how are you?");
theBuilder.Replace("Hello", "Jason");

Console.WriteLine(theBuilder); // Output: "Jason, how are you?"
```

Example: Replace the entire string

```
StringBuilder theBuilder = new("Hello, how are you?");
theBuilder.Replace(theBuilder.ToString(), "Completely new text");

Console.WriteLine(theBuilder); // Output: "Completely new text."
```

Full Example: StringBuilder

```
using System.Text;

StringBuilder theBuilder = new("Hello World!");

theBuilder.Append(" Today is going to be great!");
Console.WriteLine(theBuilder); // Output: Hello World! Today is going to be great!

theBuilder.Replace("World", "Jason");
Console.WriteLine(theBuilder); // Output: Hello Jason! Today is going to be great!

theBuilder.Replace(theBuilder.ToString(), "Completely new text");
Console.WriteLine(theBuilder); // Output: "Completely new text."
```

MINI QUIZ: STRINGBUILDER

Mini Quiz

1. StringBuilder is mutable.

    A. True
    B. False

2. When the value of a string is assigned a new value, the string object in memory is updated with the new information.

    A. True
    B. False

3. The following code would create a StringBuilder

```
StringBuilder reply = "message";
```

    A. True
    B. False

4. Would it be better to use a String or StringBuilder on text that is going to be changed 1,000 times?

    A. String
    B. StringBuilder

Mini Quiz Answers

1. A
2. B
3. B
4. B – A string would use additional memory resources.

TRY-CATCH

Try-Catch is used in scenarios that might *throw an exception*. **Exception handling** is the name used for errors in object-oriented programs, meaning if there is an exception within the **try** section of code, the program will go directly to the **catch** block of code. The catch block can specify the type of exception being looked for. In the example below, this will handle all exceptions.

Example: Try-Catch

```
try
{
  // Code that could potentially cause an exception
}
catch (Exception ex)
{
  // Handle the exception
}
```

Finally can optionally be added to the end of a Try-Catch. The code in the finally statement will *always* run. A common use for finally is if a file is open, finally can guarantee that the file will be closed even if an error occurs.

Example: Finally and multiple catches

```
try
{
  // Code that could potentially cause an exception
}
catch (OverflowException ex)
{
  // Handle the exception if it is an OverflowException
}
catch (Exception ex)
{
  // Handle all other exceptions
}
finally
{
  // This code will always execute if there is or isn't an exception.
}
```

Full Example: Try-Catch handles an exception if incorrect data is entered, such as: "asdf"

```csharp
try
{
  int theDay;
  int theMonth;
  int theYear;

  Console.WriteLine("Enter a day between 1 to 31");
  theDay = int.Parse(Console.ReadLine());

  Console.WriteLine("Enter a month between 1 to 12");
  theMonth = int.Parse(Console.ReadLine());

  Console.WriteLine("Enter a 4-digit year");
  theYear = int.Parse(Console.ReadLine());

  DateTime dt = new(theYear, theMonth, theDay);
  string displayDate = string.Format("{0:dddd, MMM d, yyyy}", dt);

  Console.WriteLine(displayDate);
}
catch (Exception ex)
{
  Console.WriteLine($"Exception: {ex.ToString()}");
}
finally
{
  // Finally can optionally be added to a Try-Catch.
}
```

MINI QUIZ: TRY-CATCH

Mini Quiz

1. There can be multiple Finally blocks in a Try.

    A. True
    B. False

2. The most generic Catch block should gdast.

    A. True
    B. False

3. The Finally block only executes if there is anexception.

    A. True
    B. False

4. The Finally block is required.

    A. True
    B. False

5. The following code will catch all exceptions.

```
catch (Exception ex)
{
    // Handle exception
}
```

    A. True
    B. False

Mini Quiz Answers

1. B
2. A
3. B
4. B
5. A

UNIT TESTING

Unit Tests verify that the logic of code is working as expected. Unit testing breaks down a program into small, testable, individual units.

Walkthrough : Setting up a .NET Console Application xUnit Test

The objective of this program is to build a calculator that adds two numbers. Unit Testing will be included to validate that the program performs as designed.

1. Create a Console Application and name it UTCalculator.

2. Create a class named "CalculatorFeatures" and add a method called "AddTwoNumbers".

```
namespace UTCalculator
{
public class CalculatorFeatures
{
   public static int AddTwoNumbers(int num1, int num2)
   {
     int result = num1 - num2;
     return result;
   }
 }
}
```

3. Add the following "Console.WriteLine" above the "namespace UTCalculator".

```
Console.WriteLine(UTCalculator.CalculatorFeatures.AddTwoNumbers(10, 4)); // Output: 6
```

The code above will run because there are no *syntax errors* in it. However, there is a *logic error* (10 + 4 should equal 14 not 6). A Unit Test will be added to catch this type of error.

4. In Solution Explorer, right click on Solution, click on Add, then click New Project.

5. In the new Project screen, search for xUnit Test Project or change the project type to Test. Click on xUnit Test Project (.NET Core) and then click Next.

6. Name it "CalculatorTest" and click Next and then Create.

The template for a xUnit Test has now been created. Next a reference needs to be added for the Unit Test to be connected to UTCalculator.

In the Solution Explorer expand CalculatorTest, right click on Dependencies, and click Add Project Reference. The Reference Manager will open. On the left menu, click on Projects -> Solution and add a checkmark next to UTCalculator. Click OK.

Make the following modifications to the Project CalculatorTest, UnitTest1.cs file.

First, add the using declaration for UTCalculator:

```
using UTCalculator; // Needed to access UTCalculator class
```

Next, modify the method below [Fact] to:

```
[Fact]
public void TestAddMethodResultShouldBeNine()
{
    int result = CalculatorFeatures.AddTwoNumbers(5, 4);
    Assert.Equal(9, result);
}
```

Unit Test names should be self-documenting. In our example, the name "TestAddMethodResultShouldBeNine" describes exactly what is expected.

Within the test "Assert.Equal(9, result)" is used to compare the expected result with the answer that the method response returns. When the test is run if 9 matches the result, it will pass the test, otherwise it will fail.

Full Example: xUnit Test

```
using Xunit;
using UTCalculator; // Needed to access UTCalculator class

namespace CalculatorTest
{
    public class UnitTest1
    {
        [Fact]
        public void TestAddMethodResultShouldBeNine()
        {
            int result = CalculatorFeatures.AddTwoNumbers(5, 4);
            Assert.Equal(9, result);
        }
    }
}
```

On the top menu bar select Test-> Run All Tests. The test will *fail*. If the Test Explorer did not come up, select Test -> Windows -> Test Explorer.

Click on the failed test for more information. A message will say "Assert.Equal() Failure, Expected: 9 Actual: 1". This indicates a logical error in the "AddTwoNumbers" method.

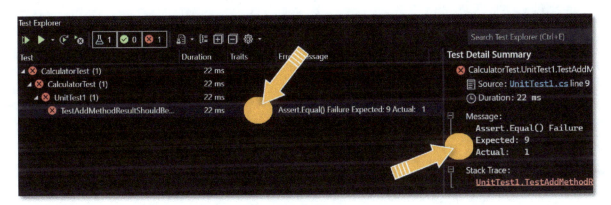

It is now clear that the numbers are being *subtracted*: "int result = num1 - num2;"

Change the minus sign to a *plus* sign: "int result = num1 + num2;"

```
public class CalculatorFeatures
{
    public static int AddTwoNumbers(int num1, int num2)
    {
        int result = num1 + num2;
        return result;
    }
}
```

Run the test again to verify all tests are passed.

UNIT TESTING: FACT & THEORY

With xUnit tests, they are either marked with the attribute [Fact] or [Theory]. **Fact** indicates a test, and **Theory** gives the ability to pass multiple data tests to the same unit test using **InlineData**. In the previous Unit Test walkthrough, the **Assert** class was utilized to compare results. The Assert class contains various methods that return a bool result. In the following examples "Assert.Equal" will be used.

Below is the class and method with which the xUnit Test will correspond. This is the same code previously used in the Unit Testing walkthrough.

```csharp
public class CalculatorFeatures
{
    public static int AddTwoNumbers(int num1, int num2)
    {
        int result = num1 + num2;
        return result;
    }
}
```

Example: Fact - Assert.Equal

```csharp
[Fact]
public void TestAddMethodResultShouldBeNine()
{
    int result = CalculatorFeatures.AddTwoNumbers(5, 4);
    Assert.Equal(9, result);
}
```

Example: Theory – Assert.Equal

```csharp
[Theory]
[InlineData(5, 4, 9)]
[InlineData(6, 1, 7)]
[InlineData(2, 3, 5)]
public void AddTwoNumbersAndGetResult(int firstNum, int secondNum, int expectedResult)
{
    int result = CalculatorFeatures.AddTwoNumbers(firstNum, secondNum);
    Assert.Equal(expectedResult, result);
}
```

DIRECTIVES

Preprocessordirectives are useful in improving readability, reducing complexity, and helping with maintenance of a program.

Example: #region and #endregion directives

- #region is used to indicate a certain block of code. This will help keep your code organized as blocks and can be expanded or collapsed in Visual Studio.

```
#region DisplayPrompts
Console.WriteLine("Hello");
Console.WriteLine("World");
#endregion
```

Full Example: #define, #if, #else, and #endif

- #define must be in the top of the code.
- The #Else statement in Visual Studio is now grayed out, indicating this section of code will not be compiled, because DevMode is defined.

```
#define DevMode

string myDatabase = "";

#if (DevMode)
{
  Console.WriteLine("In development Mode, using local database");
  myDatabase = "C:\\LocalDatabase....";
}
#else
{
  Console.WriteLine("In development Mode, using local database");
  myDatabase = "C:\\LocalDatabase...."
}
#endif
```

Below is a list of preprocessor directives:

Conditional compilation: #if, #else, #elif, #endif, #define, #undef

Other: #warning, #error, #line, #region, #endregion, #pragma, #pragma warning, #pragma checksum

WRITING/READING TEXT FILES

Variables are temporary in that once the program is turned off, the data is lost. **Writing/Reading** to a text file is a permanent way to handle data that needs to be saved when the program is turned off.

The **StreamWriter** class writes to text files. This class in contained within "System.IO" namespace To use this class, include the following at th e top of the page.

```
using System.IO; // C# 10 comes with Implicit   Usings, meaning the most popular using namespaces such
as this one are already built in so there is no need to declare it
```

A "using" statement is also used in front of StreamWriter . The using statement removes the SteamWriter from memory when it is no longer needed.

Full Example: Write to a file ; the file will be saved in the folder of your current project

```
string fileName = "test.txt" ;
string textToAdd = "Example text to save";

// true means if this file already exists, append it
using (StreamWriter writer = new(fileName, true))
{
    writer.Write(textToAdd);
    // A text file named test.txt will be added to the Project Folder Name/bin/Debug/net 6.0/test.txt
}
```

The ability to read from this file is also needed and for that use **StreamReader,** which is set up with a very similar syntax to StreamWriter.

Example: Read from a file, 1 line or whole text file

```
using (StreamReader reader = new(fileName))
{
    string onlyOneLine = reader.ReadLine();
    Console.WriteLine(onlyOneLine);
}

using (StreamReader reader = new(fileName))
{
    string wholeTextfile = reader.ReadToEnd();
    Console.WriteLine(wholeTextfile);
}
```

INTERFACES

An **interface** behaves as a contract between itself and any class where it is used. A class that implements an interface must now implement all its members. Interfaces can only contain declarations and all the members are implicitly abstract and public. Interfaces are usually named with a capital "I", although it is not required.

Example: Interface

```csharp
interface IPets
{
  void Greeting();
  string Name { get; set; }
}

public class Cat : IPets
{
  // IPets interface was used on this class. The class must now have a Greeting method and a Name
  // property, otherwise there will be an error.
  public string Name { get; set; }

  public void Greeting()
  {
    Console.WriteLine($"Hello cat named {Name}");
  }
}
```

C# does not allow multiple inheritance in classes; however, multiple interfaces are allowed. If interfaces are added to a class that is also inheriting from a base class, the interfaces must follow the base class.

Example: Multiple interfaces and base class

```csharp
public class DeckOfCards : ABaseClass, IInterface, IAnotherInterface
{
  // Empty class
}
```

Full Example: Interface

```csharp
Cake exampleCake = new();
Pasta examplePasta = new();

// Cake and Pasta classes both use the IFood interface so both can be stored in a List of IFood
List<IFood> foodList = new()
{
  exampleCake,
  examplePasta
};

foreach (IFood food in foodList)
{
    food.Prepare();
}

interface IFood
{
  void Prepare();
}

public class Cake : IFood
{
    public void Prepare()
    {
      Console.WriteLine("Bake cake 20 mins");
      Console.WriteLine("Frost Cake");
    }
}

public class Pasta : IFood
{
    public void Prepare()
    {
      Console.WriteLine("Boil pasta for 12 minutes");
      Console.WriteLine("Strain water and add sauce");
    }
}
```

There are useful built-in interfaces, such as **IEnumerable**, **IList**, **IDictionary**, and **IComparable**. Adding the interface IEnumerable to a class means it will iterate. The class must also now contain **IEnumerator**. In the example below, a class for a deck of cards is created and iterated over.

Full Example: IEnumerable and IEnumerator

```csharp
using System.Collections;

DeckOfCards theDeck = new();

foreach (Card itemCard in theDeck)
{
  Console.WriteLine($"Card: {itemCard.Suit} {itemCard.Name} ({itemCard.Value})");
}

class Card
{ // Set up the properties of a playing card
  public string Name { get; set; }
  public int Value { get; set; }
  public string Suit { get; set; }

  public Card(string name, int value, string suit)
  {
    Name = name;
    Value = value;
    Suit = suit;
  }
}

class DeckOfCards : IEnumerable // This class can now be iterated over to display a deck of cards
{
  private List<Card> deckList = new();

  public DeckOfCards()
  {
    deckList.Add(new Card("Two", 2, "Spade"));
    deckList.Add(new Card("Three", 3, "Spade"));
    deckList.Add(new Card("Four", 4, "Spade"));
  }
  // IEnumerator must be used, because DeckOfCards inherits IEnumerable
  public IEnumerator GetEnumerator()
  {
    return deckList.GetEnumerator();
  }
}
```

MINI QUIZ: INTERFACES

Mini Quiz

1. A class can inherit from multiple interfaces.

    A. True
    B. False

2. The following is valid code.

```
public class ExampleClass: IInterfaceOne, IInterfaceTwo, ABaseClass
{
  // Empty class
}
```

    A. True
    B. False

3. Classes that inherit from an interface must implement all of the interface items.

    A. True
    B. False

4. A List data type can be an Interface.

    A. True
    B. False

Mini Quiz Answers

1. A
2. B – The two interfaces must be after the class "ABaseClass"
3. A
4. A

YIELD

Yield return and **yield break** are used when implementing an iterator (IEnumerable). The *yield return* statement returns the next element in the sequence, whereas *yield break* ends the iteration.

Full Example: Yield return and yield break

```
Console.WriteLine( "Output all numbers greater than 5, stop if number is 7");

List<int> MyNumbers = new() { 9, 4, 20, 3, 7, 12 };

foreach (int item in GreaterThan5StopIf7())
{
  Console.Write( $"{item} "); // Output: 9 20
}

IEnumerable<int> GreaterThan5StopIf7()
{
  foreach (int item in MyNumbers)
  {
    if (item == 7)
    {
      yield break;        //  End iteration
    }
    else if (item > 5)
    {
      yield return item; // Return element and continue iteration
    }
  }
}
```

DELEGATES

Delegates are a variable that can store a method and can then be passed around as needed. Previously the variables discussed held data such as strings or objects, however methods can be treated the same way. In the example below, a delegate named "MathExample" can store methods that accept two integer values and return a string value.

Example: Create a delegate

```
public delegate string MathExample(int num1, int num2);
```

Example: Use a delegate

```
MathExample calculateMath = AddNumbers; // Store method AddNumbers
```

Full Example: Delegate

```
int number1 = 3;
int number2 = 7;

MathExample calculateMath = AddNumbers;          // Store method AddNumbers
Console.WriteLine(calculateMath(number1, number2)); // Output: 3 + 7 = 10

static string AddNumbers(int a, int b)
{
   return $"{a} + {b} = {a + b}";
}

public delegate string MathExample(int num1, int num2);
```

Example: Declare and instantiate an array of delegates

```
MathExample[] mathOperations = { AddNumbers, MultiplyNumbers };
```

Full Example: Array of delegates

```csharp
int number1 = 3;
int number2 = 7;

// Declare and instantiate an array of MathExample
MathExample[] mathOperations = { AddNumbers, MultiplyNumbers };

foreach (var mathOperation in mathOperations)
{
  Console.WriteLine(mathOperation(number1, number2));
}
 // Output: 3 + 7 = 10
 // Output: 3 * 7 = 21

static string AddNumbers(int a, int b)
{
  return $"{a} + {b} = {a + b}";
}

static string MultiplyNumbers(int a, int b)
{
  return $"{a} * {b} = {a * b}";
}

public delegate string MathExample(int num1, int num2);
```

MINI QUIZ: DELEGATES

Mini Quiz

1. A delegate can store a method.

   A. True
   B. False

2. The following code is valid way to declare a delegate.

```
public delegateint Word(string text);
```

   A. True
   B. False

3. You can notstore a delegate in an array.

   A. True
   B. False

4. When creating a delegate there can be only one parameter.

   A. True
   B. False

Mini Quiz Answers

1. A
2. A
3. B
4. B

EVENTS

Events allow classes to notify other classes when something occurs. Graphical User Interfaces (GUI) applications often use Events to indicate when something occurs. One GUI example is when a button is pressed and another class is *listening* to handle that event.

There are multiple components to an event. Each of the examples below cover one of these components. Following all the examples is a full example that has all the components combined.

An event needs to be defined inside a class. Using the "event" keyword enables others to attach to this event. Specify the delegate type for the type of methods that can be attached to the event. In the example below, a predefined delegate called "EventHandler" is used. The EventHandler delegate returns void and has two parameters (object, EventArgs). Object is the *sender* (what sent the event) and EventArgs object (contains basic information about the event).

Example: Define an event

```
public event EventHandler WelcomeChanged; // Define the event
```

The code below verifies that an event handler is attached to an event. Methods that *raise* events typically start with the word "On" .

Example: Raise an event

```
public void OnWelcomeChanged()
{
   WelcomeChanged?.Invoke(this, EventArgs.Empty);
}
```

In the example below, when "TheMessage" value is set, then it will call the method "OnWelcomeChanged" to raise the event.

Example: Call method to raise an event

```
private string theMessage;
public string TheMessage
{
 get { return theMessage; }
 set {
     theMessage = $"Hello, {value}";
     OnWelcomeChanged(); // Call OnWelcomeChanged when the value is changed
   }
}
```

The method below handles what to do when the change is detected. This method needs to be set up, based on the delegate used, in this case the "EventHandler" delegate.

Example: Handle the event

```csharp
public void HandleWelcomeChanged(object sender, EventArgs eventArgs)
{
    Console.WriteLine( "The welcome message has changed!!");
}
```

Below is how to attach an event handler to an event, the "+=" operator is used to accomplish this. You can attach multiple event handlers to an event if needed.

Example: Attach event handler to an event

```csharp
welcomeMessage.WelcomeChanged += welcomeMessage.HandleWelcomeChanged;
```

Full Example: Event

```csharp
Greetings welcomeMessage = new();

// Attach event handler to event
welcomeMessage.WelcomeChanged += welcomeMessage.HandleWelcomeChanged;

welcomeMessage.TheMessage = "Adam";
Console.WriteLine(welcomeMessage.TheMessage);

class Greetings
{
  private string theMessage;
  public string TheMessage
  {
    get
    {
      return theMessage;
    }
    set
    {
      theMessage = $"Hello, {value}";
      OnWelcomeChanged(); // Call OnWelcomeChanged when the value is changed
    }
  }

  public event EventHandler WelcomeChanged; // Define the event
  // EventHandler is a predefined delegate that returns void and has 2 parameters.
  // The first parameter is an object and the second is an EventArgs object

  public void OnWelcomeChanged() // Methods that raise events usually start with the "On"
  {
    // The code below raises the event if the event is not null (verifies an event handler
    // is attached to the event)
    WelcomeChanged?.Invoke(this, EventArgs.Empty);
  }

  // This method handles what to do when the change is detected
  public void HandleWelcomeChanged(object sender, EventArgs eventArgs)
  {
    Console.WriteLine("The welcome message has changed!!");
  }
}
```

MINI QUIZ: EVENTS

Mini Quiz

1. Events notify other classes when something occurs.

    A. True
    B. False

2. Events are oftenused in a GUI.

    A. True
    B. False

3. Events only have one component.

    A. True
    B. False

4. The "+=" operator attaches an event handler to an event.

    A. True
    B. False

Mini Quiz Answers

4. A
3. B
2. A
1. A

LAMBDAS

A **lambda** is essentially a method without a declaration; it uses clear and short syntax. A lambda can be written as a shorthand method directly in the place it is intended to be used. This is especially useful on small methods that are used only once. The lambda operator is "=>" and can be read as "goes to". The lambda operator separates the expression into two parts (left side parameter, right side lambda body).

Example: Lambda

```
ANewNumber addFive = x => x + 5;
```

Full Example: Lambda using a delegate

```
ANewNumber addFive = x => x + 5;
Console.WriteLine(addFive(2)); // Output: "7"

delegate int ANewNumber(int input);
```

Full Example: Lambdas and lists

```
List<int> numbers = new() { 5, 3, 4, 5, 6, 7 };

// Gets all 5's from list
List<int> numberFives = numbers.FindAll(x => x == 5);
Console.WriteLine($"There are {numberFives.Count} 5's in the list");

// Gets all odd numbers from list
List<int> oddNumbers = numbers.FindAll(x => (x % 2) != 0);
Console.WriteLine($"There are {oddNumbers.Count} ODD numbers:");
```

The lambda expression syntax can also be used on various *expression-bodied members* that are a single expression. In the example below, a single expression method is simplified using the lambda expression syntax.

Full Example: Expression-bodied members with lambda expression syntax

```
Console.WriteLine(SquareNumber1(5));  // Output: 25
Console.WriteLine(SquareNumber2(5));  // Output: 25

static int SquareNumber1(int number) // "Regular" Method
{
   return number * number;
}
//  The same method but using lambda expression syntax
static int SquareNumber2(int number) => number * number;
```

MINI QUIZ: LAMBDAS

Mini Quiz

1. The following is a lambda operator "<=".

   A. True
   B. False

2. The lambda operator separates the expression into two parts.

   A. True
   B. False

3. The following code would output 7.

```
Console.WriteLine(AMethod(2));
static int AMethod(int number) => number + number + number + 1;
```

   A. True
   B. False

Mini Quiz Answers

THREADS

Threads add the functionality of running multiple sections of code simultaneously. Threading allows the code to run on *multiple* processors. This provides a performance boost.

To use threads, include the following at the top of the page:

```
using System.Threading; // C # 10 comes with Implicit Usings, meaning the most popular using namespaces such as this one are already built in so there is no need to declare it
```

The "Thread" class constructor used below contains a delegate that accepts methods that return void; only methods that return void can be used.

Example: Create a thread object

```
Thread aThread = new(CountTo200);
```

In the example below, the "Start" method is used and it can accept zero or one parameter.

Example: Start a thread

```
aThread.Start();    // 0 parameters
bThread.Start(45); // 1 parameter
```

Full Example: Thread – No parameters

```
Thread aThread = new(CountTo200);
Thread bThread = new(CountTo200);
aThread.Start();
bThread.Start();

static void CountTo200()
{
  for (int i = 1; i <= 200; i++)
  {
    Console.WriteLine(i);
  }
}
```

The previous example uses the "Start" method to create two threads that will begin executing the method "CountTo200". At this point the two threads are now running simultaneously. The results will not output in sequence 1-200 each time. Both threads are running at the same time and will each output their numbers as their thread is being run. For example, the results might be 1-70 and then the other thread could output 1-60, then the first thread could output 71-120 and so on.

In the example below, the "Join" method instructs the original thread to wait until "aThread" has completed. The original thread will be "frozen" until it finishes, essentially joining the two threads.

Example: Join

```
aThread.Join();
```

Full Example: Join

```
Thread aThread = new Thread(CountTo200);
Thread bThread = new Thread(CountTo200);
aThread.Start();
aThread.Join(); // Wait until aThread is finished
bThread.Start();
bThread.Join(); // Wait until bThread is finished

static void CountTo200()
{
  for (int i = 1; i <= 200; i++)
  {
    Console.WriteLine(i);
  }
}
```

The example below demonstrates using a parameter in the "Start" method. It is important to note that the method parameter must be an *object*. The object can later be cast In the example below, it is cast from an object to an int.

Full Example: Thread – One parameter

```
Thread aThread = new Thread(CountTo);
aThread.Start(45); // 1 parameter

static void CountTo(object count)
{
  for (int i = 1; i <= (int)count; i++)
  {
    Console.WriteLine(i);
  }
}
```

ACTIVITY: THREADS

Activity: This activity focuses on using threads to display which thread is being used. Create a Console Application for Threads App that meets the two requirements listed below.

1. Create 3 threads.

2. Create a method that counts to 100 and displays which thread is running.

The finished output of the program should look like the example below (results will vary) :

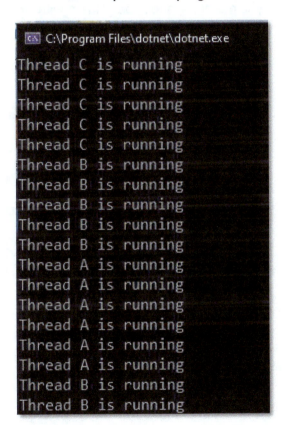

Need help with the solution? https://www.unQbd.com/Solutions/CSharp5th/Threads

ASYNCHRONOUS

Asynchronous programming helps enhance the responsiveness of applications by using non-blocking operations to prevent bottlenecks that could slow down or freeze an application. The two main components to asynchronous programming are the keyword modifiers **async** and **await**.

A method using the async modifier enables the use of the await operator, which now must be included at least once within the method. When the await operator is reached, the original caller method will continue and the async method will continue to process until it is completed.

Async methods must have a return type of void, **Task, Task<T>**, or any other type that has a *GetAwaiter* method. The naming convention for async methods is to append them with an "Async" suffix.

Full Example: Async void method

```
FirstMethodAsync();
Console.WriteLine( "System is not Frozen');
Console.ReadLine();

static async void FirstMethodAsync()
{
  Console.WriteLine( "Task Started");
  await Task.Delay(3000); // Delays the method for 3 seconds
  Console.WriteLine( "Task Finished");
}
 // Output: "Task Started"
 //          "System is not Frozen"
 //          "Task Finished"
```

In the example above, the method "FirstMethodAsync()" contains a delay that would generally "freeze" a program for 3 seconds. Because this is an async method, when await is reached the original caller of the method will continue *and* the async method will keep processing.

Async methods can return with void as demonstrated previously, however they generally return with Task or Task<TResult>, which encapsulates information. Task is used if there is no return statement, and Task<TResult> is used if there is a return statement.

To use tasks, include the following at the top of the page:

```
using System.Threading.Tasks; // C# 10 comes with Implicit Usings, meaning the most popular using namespaces such as this one are already built in so there is no need to declare it
```

When the task return type is used instead of void, the method call can now use await.

Example: Await method

```
await SecondMethodAsync();
```

Full Example: Async Task method

```
FirstMethodAsync();
Console.WriteLine( "System is not Frozen');
Console.ReadLine();

static async void FirstMethodAsync()
{
  Console.WriteLine( "Task Started");
  await SecondMethodAsync();
  Console.WriteLine( "First Task Finished");
}

static async Task SecondMethodAsync()
{
  await Task.Delay(3000); // Delays the method for 3 seconds
  Console.WriteLine( "Second Task Finished');
}
 // Output: Task Started
 //          System is not Frozen
 //          Second Task Finished
 //          First Task Finished
```

Full Example: Async Task<TResult> method

```
FirstMethodAsync();
Console.WriteLine("System is not Frozen");
Console.ReadLine();

static async void FirstMethodAsync()
{
  Console.WriteLine("Task Started");
  Console.WriteLine(await SecondMethodAsync());
  Console.WriteLine("First Task Finished");
}

static async Task<string> SecondMethodAsync()
{
  await Task.Delay(3000); // Delays the method for 3 seconds
  return ("Second Task Finished");
}
 // Output: Task Started
 //          System is not Frozen
 //          Second Task Finished
 //          First Task Finished
```

It is very common to see async programming on graphical applications. For example, if a button is pressed and content from the internet needs to be downloaded, the whole program should not be frozen while it downloads. The program should download the content in the background while the user is still able to use the application.

Mini Quiz

1. Asynchronous programing helps reduce the chance of a program responsiveness from "freezing".

    A. True
    B. False

2. The naming convention for an async method is to append them with an "async" prefix.

    A. True
    B. False

3. If the async modifier is used, the await operator also has to be used.

    A. True
    B. False

4. Methods that are async must have a return type that uses a GetAwaiter method.

    A. True
    B. False

Mini Quiz Answers

1. A
2. B – Append suffix with Async not prefix
3. A
4. A

QUERY EXPRESSIONS: LINQ – QUERY SYNTAX

A **query expression** statement extracts specific information from a collection of data. The query syntax in C# is called "Language Integrated Query", commonly referred to by its shortened name, "**LINQ** ".

LINQ queries are used on collections of data, such as Lists or Arrays, which are container types for the "IEnumerable<T>" interface. Any container type of "IEnumerable<T>" can use a LINQ query.

To use a LINQ query, include the following at the top of the page:

```
using System.Linq; // C# 10 comes with Implicit Usings, meaning the most popular using namespaces
such as this one are already built in so there is no need to declare it
```

In the examples below, the various *clauses* that make up a query are covered. Clauses are a smaller syntax than a full expression; LINQ queries require multiple clauses to be a complete expression.

Queries always start with a "from" clause. The from clause defines a source of information for the qu ery and a *range variable* . A range variable is a local variable that is available throughout the query, like the variable in a "ForEach" loop.

A full example at the end of this section contains all the classes and lists to make the following examples functional.

Example: "From" clause

```
from o in students
```

The select clause specifies the part or whole object to be used for the results of the query expression. The example below demonstrates selecting the entire object.

Example: "Select" clause

```
select o;
```

The results from a LINQ query are always output as "IEnumerable<T>" and can be converted to a list or array if needed using "ToList()" or "ToArray()" .

Example: Full expression - assign results to a variable

```
IEnumerable<StudentInfo> aStudents = from o in students
                                     select o;
```

Queries can be created using only "from" and "select" clauses; the "where" clause specifies which items to use.

Example: "Where" clause

```
where o.Grade > 90
```

Example: Full expression using "where" clause

```
IEnumerable<StudentInfo> aStudents = from o in students
                                     where o.Grade > 90
                                     select o;
```

The example below uses the "let" clause which creates *derived range variable s*. A derived range variable uses the range variable created in the "from" clause. "Let" clauses define a variable to use in multiple places in the query or to simplify a complex operation. In the example below, "select" also selects a string instead of an object, and IEnumerable is modified to reflect this change.

Example: Full expression using "let" clause

```
IEnumerable<string> studentAverages = from o in students
                                      let averageGrade = (o.Grade1 + o.Grade2) / 2
                                      select $"{o.Name} grade average {averageGrade}";
```

A "join" clause combines two collections together based on a condition. In the example below, the collections are joined together when the "StudentID" values are *equal* in both collections. The left side of the equal sign must reference an earlier range variable and the right side references a new range variable; this is called an **inner join**.

Example: Full expression using "join" clause

```
var inUseBooks = from o in students
                 join b in books on o.StudentID equals b.StudentID // Inner join
                 select new { o.Name, b.BookName };                // New collection using these types
```

The "OrderBy" clause sorts all items in a collection. The default is ascending, however they can also be specified for descending.

Example: "OrderBy" clause

```
orderby o.Name                          // Ascending by default
orderby o.Name descending               // Descending example
orderby o.Name, o.StudentID descending  // Multilevel sort: Sort by name then StudentID
```

Example: Full expression using "OrderBy" clause

```
IEnumerable<StudentInfo> aStudents = from o in students
                                     orderby o.Name // Ascending by default
                                     select o;
```

The "group" clause bundles elements into groups based on the information specified. In the example below, the books collection will have a group for each "BookName". Previously the "select" clause had been needed to finish a full expression, however the "group" clause is the other way to finish a full expression.

Example: Full expression using "group" clause

```
var bookGroups = from o in books
                 group o by o.BookName;
```

In the example above, "var" is representing the syntax "IEnumerable<IGrouping<TKey, TElement>>". "TKey" in this case is going to be a string for the book name and "TElement" is the class from which the collection was made.

Full Example: Query Expressions

```csharp
// Lists, students, and books will be used as our collections for this example
List<StudentInfo> students = new()
{
  new StudentInfo {StudentID = 1, Name = "Jonathan", Grade1 = 95, Grade2 = 90},
  new StudentInfo {StudentID = 2, Name = "Maria", Grade1 = 92, Grade2 = 85},
  new StudentInfo {StudentID = 3, Name = "Marcos", Grade1 = 81, Grade2 = 91}
};

List<BookInfo> books = new()
{
  new BookInfo {StudentID = 1, BookName = "C# Fundamentals"},
  new BookInfo {StudentID = 1, BookName = "Microsoft Magazine"},
  new BookInfo {StudentID = 2, BookName = "C# Fundamentals"}
};

// Minimal query expression
IEnumerable<StudentInfo> students1 = from o in students
                                     select o;

// Where specifies which items to use, in this example all Grade1 greater than 90
IEnumerable<StudentInfo> students2 = from o in students
                                     where o.Grade1 > 90
                                     select o;

// Display grades greater than 90
foreach (var student in students2)
{
  Console.WriteLine($"Name: {student.Name} Grade: {student.Grade1}");
}
// Output: Name: Jonathan Grade: 95
//         Name: Maria Grade: 92

// Let creates derived range variables
IEnumerable<string> studentAverages = from o in students
                                      let averageGrade = (o.Grade1 + o.Grade2) / 2
                                      select $"{o.Name} grade average: {averageGrade}";

// Join combines two collections together based on a condition
var inUseBooks = from o in students
                 join b in books on o.StudentID equals b.StudentID // Inner join
                 select new { o.Name, b.BookName }; // New collection using these types

// Display books that are checked out
foreach (var inUseBook in inUseBooks)
{
  Console.WriteLine($"{inUseBook.Name} checked out book: {inUseBook.BookName}");
}

 // Output: Jonathan checked out book: C# Fundamentals
 // Output: Jonathan checked out book: Microsoft Magazine
 // Output: Maria checked out book: C# Fundamentals
```

```csharp
// OrderBy sorts all items in a collection
IEnumerable<StudentInfo> students3 = from o in students
                                     orderby o.Name // Ascending by default
                                     select o;


// Groups bundle elements into groups based on the information specified
var bookGroups = from o in books
                 group o by o.BookName;

// Display information about the groups
foreach (var bookGroup in bookGroups)
{
  Console.WriteLine($"Group Key: {bookGroup.Key}");

  foreach (var o in bookGroup)
  {
    Console.WriteLine($"Book: {o.BookName} is being borrowed by studentID: {o.StudentID}");
  }
  Console.WriteLine();
}
// Output: Group Key: C# Fundamentals
//         Book: C# Fundamentals is being borrowed by studentID: 1
//         Book: C# Fundamentals is being borrowed by studentID: 2
//
//         Group Key: Microsoft Magazine
//         Book: Microsoft Magazine is being borrowed by studentID: 1

class StudentInfo
{
  public int StudentID { get; set; }
  public string Name { get; set; }
  public int Grade1 { get; set; }
  public int Grade2 { get; set; }
}

class BookInfo
{
  public int StudentID { get; set; } = 0;
  public string BookName { get; set; }
}
```

QUERY EXPRESSIONS: LINQ – METHOD CALL SYNTAX

Method call syntax provides the same results as the query syntax covered in the previous section. All the clauses previously covered are also available through these simple method invocations that use lambda expressions.

Example: Where clause

```
// Previous query syntax
IEnumerable<StudentInfo> students2 = from o in students
                                     where o.Grade1 > 90
                                     select o;

// Method call syntax
IEnumerable<StudentInfo> students3 = students.Where(o => o.Grade1 > 90);
```

Example: Where and select clause

```
// Previous query syntax
IEnumerable<string> students2 = from o in students
                                where o.Grade1 > 90
                                select o.Name;

// Method call syntax
IEnumerable<string> students3 = students.Where(o => o.Grade1 > 90).Select(o => o.Name);
```

Example: OrderBy and select clause

```
// Previous query syntax
IEnumerable<string> students2 = from o in students
                                orderby o.Name
                                select o.Name;

// Method call syntax
IEnumerable<string> students3 = students.OrderBy(o => o.Name).Select(o => o.Name);
```

WALKTHROUGH: DATABASE

Databases store a collection of information in an organized way that can easily be added, deleted, and updated. In the following example, a local **SQL database** is created to demonstrate **CRUD** (Create, Read, Update, Delete) in a database.

Walkthrough : Creating a database

1. Create a .NET Console Application in Visual Studio and name it "StudentDB".

2. Verify that Entity Framework is installed. Right click on the project "StudentDB" and select "Manage NuGet Packages". Click the "Browse" tab.

a. Browse for "Microsoft.EntityFrameworkCore" and install if it is not installed.

b. Browse for "Microsoft.EntityFrameworkCore.Tools" and install if it is not installed.

c. Browse for "Microsoft.EntityFrameworkCore.SQLServer" and install if it is not installed.

3. Create a folder to store and setup the model by doing the following.

    a.   Right click on the project "StudentDB" and select Add -> New Folder. Name it "People"
    b.   Right click on the People folder and add a Class called Student

4. Add the following code to the Student class to create the model.

```
class Student
{
  public int Id { get; set; }
  public string Name { get; set; }
  public int Grade { get; set; }
}
```

5. Create an additional folder called "Data" for the project and a class in the Data folder called "StudentDBContext". This class will manage the connection to the database.

6. The class StudentDBContext needs to inherit from the Entity Framework class DBContext.

a. Update the class to "class StudentDBContext : DbContext"

b. To use DBContext, add "using Microsoft.EntityFrameworkCore;" to the top of the page

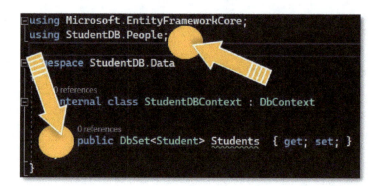

7. To handle the information from the Student class model previously created, the DbSet is used "public DbSet <Student> Students { get; set; }". An additional using statement "using StudentDB.People;" is also added to access the Student class.

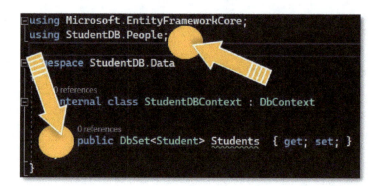

8. A connection string is added to connect to the database. The class should now look like the example below.

```
Internal class StudentDBContext : DbContext
{
    public DbSet<Student> Students { get; set; }

    // Connect to database
    protected override void OnConfiguring(DbContextOptionsBuilder optionsBuilder)
    {
optionsBuilder.UseSqlServer(@"Server=(localdb)\MSSQLLocalDB;Database=StudentDatabase;Trusted_
Connection=True;");
    }
}
```

9. Use migration to create the database. The purpose of migration is to keep the database in sync with the models.

a. On the top menu click on Tools -> Nuget Package Manager -> Package Manager Console

b. In the Package Manager Console input "Add-Migration InitialMigration". This scaffolds the initial set of tables for the model. The database has not been created yet, but this is the schema it will use

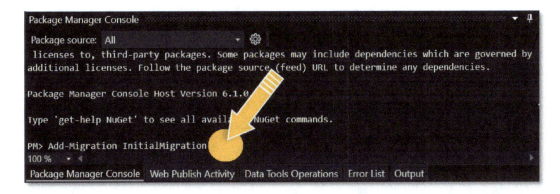

c. After pressing enter, notice that a new folder called "Migrations" was added to the project. This folder contains the schema and other information the database will need

d. Input "Update-Database" in the Package Manager Console. The database is not created yet; this will create the database

10. View the database by using SQL Server Object Explorer. If it is not currently on the left side of Visual Studio, then click on View -> SQL Server Object Explore. The database will be located at SQL Server -> (localdb).. -> Databases -> StudentDatabase.

11. View the table created by expanding the StudentDatabase -> Tables and then right click on dbo.Students -> View Data.

12. It is now possible to input data directly into the database table. Enter in a Name and Grade. Make sure the Grade is a number. Notice a new row is created when data starts to be input and multiple records can be entered. However, data is not usually input directly into the database. In the next section ("Database"), how to add input using code is covered.

DATABASE

The examples below continue the previous examples from "Walkthrough: Database".

Below is an example of adding items to a database by code. Adding this code snippet to the Program.cs Main method will add the two students every time the application is loaded. For the code snippet below to work, add two using statements "using StudentDB.Data;" and "using StudentDB.People;" to the top of the page.

Example: Adding items to a database

```csharp
using (var db = new StudentDBContext())
{
    // Add the following students to Database
    db.Students.AddRange(new Student
                        {
                          Name = "Wyatt",
                          Grade = 95
                        },
                        new Student
                        {
                          Name = "Kristen",
                          Grade = 98
                        });

    // Save the changes
    db.SaveChanges();
}
```

In the example below, a "ForEach" loop is used to loop through an entire database table and display the results.

Example: Display results

```csharp
using (var db = new StudentDBContext())
{
    foreach (var theStudent in db.Students)
    {
        Console.WriteLine($"Name: {theStudent.Name} Grade: {theStudent.Grade}");
    }
}
```

The example below will query the database table using the **where** keyword. When the "Where" keyword is used, only results that match the search parameters will be returned.

Example: Where query and display results

```
using (var db = new StudentDBContext())
{
  // Search for students named "Wyatt"
  // To use the query Where below, use "Using System.Linq"
  var studentInfo = db.Students.Where(x => x.Name == "Wyatt");

  foreach (var theStudent in studentInfo)
  {
    Console.WriteLine($"Name: {theStudent.Name} Grade: {theStudent.Grade}");
  }
}
```

Using the **contains** keyword can provide flexibility to a search. For example, if a user searches "Ale", then the results will contain names that contain "Ale" such as "Alex" or "Alessandra".

Example: Contains

```
using (var db = new StudentDBContext())
{
  var query = db.Students.Where(s => s.Name.Contains("Ale"));
}
```

Using the **OrderBy** keyword sorts the results numerically and alphabetically.

Example: OrderBy

```
using (var db = new StudentDBContext())
{
  var query = db.Students.OrderBy(s => s.Name);
}
```

Other popular keywords are **Average, Sum, Max**, and **Min**. An example on how to get the Max value is below.

Example: Max value

```
using (var db = new StudentDBContext())
{
  var query = db.Students.Max(s => s.Age);
}
```

To remove items from a table, use **RemoveRange** or **Remove**. After removing the items, **SaveChanges** must be called to update the database.

Example: Remove and save changes

```
using (var db = new StudentDBContext())
{
  // Example 1: Remove students with RemoveRange and the Where query
  db.Students.RemoveRange(db.Students.Where(x => x.Name == "Wyatt"));

  // Example 2: Remove without needing Linq
  foreach (var DeleteStudent in db.Students)
  {
    if (DeleteStudent.Name == "Matt")
    {
      db.Students.Remove(DeleteStudent);
    }
  }

// Save the changes
db.SaveChanges();
}
```

Use the **select** keyword to specify which columns are needed. This improves the speed of the results returned.

Example: Select columns

```
using (var db = new StudentDBContext())
{
  var query = db.Students
              .Select(s => new { s.Name });
}
```

Example: Where and Select multiple columns

```
using (var db = new StudentDBContext())
{
  var query = db.Students
              .Where(s => s.Name == "Bill")
              .Select(s => new { s.Name, s.Grade });
}
```

There are several ways to get only one result from a collection:

- **Single** should be used when there is exactly one result from a query An exception will be thrown if no results or multiple results are returned
- **SingleOrDefault** should be used the same as Single except a null value will not throw an error
- **First** should be used to display the first result. An exception will be thrown if no results are returned
- **FirstOrDefault** should be used the same as First except a null value will not throw an error

Example: Single, SingleOrDefault, First, FirstOrDefault

```
using (var db = new StudentDBContext())
{
  var querySingle = db.Students.Single(s => s.Name == "Wyatt" );
  var querySingleOrDefault = db.Students.SingleOrDefault(s => s.Name == "Wyatt" );
  var queryFirst = db.Students.First(s => s.Name == "Wyatt" );
  var queryFirstOrDefault = db.Students.FirstOrDefault (s => s.Name == "Wyatt" );
}
```

Using the **join** keyword combines information from two separate tables based on a common attribute between the two. For example, table 1 contains a student with ID "5", Name "Cesar", and Grade "98". Table 2 contains StudentID "5" and age "25". If ID matches StudentID, put the information together (Name, Grade, and Age).

Example: Join multiple tables and display results

```
using (var db = new StudentDBContext())
{
  var query = (from a in db.Students        // Database Table 1
               join b in db.Students2       // Database Table 2
               on a.ID equals b.StudentID   // The two tables will need to have something in
               select new                   // common to compare; an ID is often used
               {
                 a.Name,
                 a.Grade,
                 b.Age
               });
}
foreach (var aStudent in query)
{
  Console.WriteLine( $"Name: {aStudent.Name} Grade: {aStudent.Grade} Age: {aStudent.Age}"
}
```

FINAL THOUGHT

Congratulations on finishing the book! This book has given you the knowledge to start building a wide varies of applications. Below are a couple of tips on where to go from here.

Where to go from here:

- Continue reading programming books at www.unQbd.com.
- Practice, Practice, Practice! The best way to get more proficient at C# is to practice! Think of a personal project you would like to build and build it!

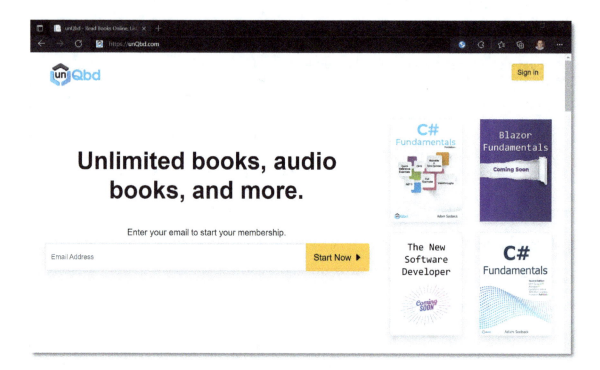